Putting Information First

☆METAPHILOSOPHY

METAPHILOSOPHY SERIES IN PHILOSOPHY

Series Editors Armen T. Marsoobian, Brian J. Huschle,
and Eric Cavallero

Putting Information First

Luciano Floridi and the Philosophy of Information

Edited by

Patrick Allo

A John Wiley & Sons, Ltd., Publication

This edition first published 2010
Originally published as volume 41, no. 3 (April 2010) of *Metaphilosophy*
Chapters © 2010 The Authors
Book compilation © 2010 by Blackwell Publishing Ltd and Metaphilosophy LLC

Blackwell Publishing was acquired by John Wiley & Sons in February 2007. Blackwell's publishing program has been merged with Wiley's global Scientific, Technical, and Medical business to form Wiley-Blackwell.

Registered Office
John Wiley & Sons Ltd, The Atrium, Southern Gate, Chichester, West Sussex, PO19 8SQ, United Kingdom

Editorial Offices
350 Main Street, Malden, MA 02148-5020, USA
9600 Garsington Road, Oxford, OX4 2DQ, UK
The Atrium, Southern Gate, Chichester, West Sussex, PO19 8SQ, UK

For details of our global editorial offices, for customer services, and for information about how to apply for permission to reuse the copyright material in this book please see our website at www.wiley.com/wiley-blackwell.

The right of Patrick Allo to be identified as the author of the editorial material in this work has been asserted in accordance with the UK Copyright, Designs and Patents Act 1988.

Wiley also publishes its books in a variety of electronic formats. Some content that appears in print may not be available in electronic books.

Designations used by companies to distinguish their products are often claimed as trademarks. All brand names and product names used in this book are trade names, service marks, trademarks or registered trademarks of their respective owners. The publisher is not associated with any product or vendor mentioned in this book. This publication is designed to provide accurate and authoritative information in regard to the subject matter covered. It is sold on the understanding that the publisher is not engaged in rendering professional services. If professional advice or other expert assistance is required, the services of a competent professional should be sought.

Library of Congress Cataloging-in-Publication data is available for this book.

9781444338676 (paperback)

A catalogue record for this book is available from the British Library.

Set in 10pt Times
By Macmillan India Ltd
Printed in Malaysia by Ho Printing (M) Sdn Bhd

01 2010

CONTENTS

NOTES ON CONTRIBUTORS

Fred Adams is professor and chair of Linguistics and Cognitive Science and professor of philosophy at the University of Delaware. He has more than a hundred publications in epistemology, philosophy of language, philosophy of mind, and cognitive science. Among his recent publications is his book *Bounds of Cognition* (with Ken Aizawa; Wiley-Blackwell, 2008).

Patrick Allo is a postdoctoral fellow of the Research Foundation—Flanders (FWO) who is based at the Centre for Logic and Philosophy of Science at the Vrije Universiteit Brussel in Belgium, and he is also affiliated with the Information Ethics Group (IEG) at the University of Oxford. His research is on the interface between philosophy of logic, formal epistemology, and philosophy of information.

Selmer Bringsjord is professor of cognitive science, computer science, logic, and management at Rensselaer Polytechnic Institute and Director of the Rensselaer AI and Reasoning Laboratory. He specializes in the logico-mathematical and philosophical foundations of AI and cognitive science, and in building AI systems. His books include *Superminds* (Kluwer), *What Robots Can and Can't Be* (Kluwer), *Artificial Intelligence and Literary Creativity* (with Dave Ferrucci, Erlbaum), *Abortion* (Hackett), and *Soft Wars* (Penguin), a novel.

Otávio Bueno is professor of philosophy at the University of Miami. His work focuses on the philosophies of science, mathematics, and logic. He has published articles in *Noûs, Mind, Philosophy of Science, Synthese, Australasian Journal of Philosophy, Erkenntnis, Analysis, Journal of Philosophical Logic,* and *Philosophia Mathematica.* He edited, with Øystein Linnebo, *New Waves in Philosophy of Mathematics* (Palgrave Macmillan, 2009). His work has been funded by the National Science Foundation, among other institutions.

Terrell Ward Bynum is director of the Research Center on Computing and Society and professor of philosophy at Southern Connecticut State University. He has been chair of the Committee on Philosophy and Computers of the American Philosophical Association, as well as chair of the Committee on Professional Ethics of the Association for Computing Machinery. He is a recipient of the Barwise Prize of the APA, the Weisenbaum Award of the International Association for Ethics and Information Technology, and the Covey Award of the International Association for Computing and Philosophy.

Timothy R. Colburn is associate professor of computer science at the University of Minnesota—Duluth. He received his Ph.D. in philosophy from Brown University and his M.S. in computer science from Michigan State University. He has worked as a philosophy and computer science professor, a software engineer, and a research scientist in artificial intelligence for the aerospace and defense industries. He is the author of *Philosophy and Computer Science* and editor of *Program Verification: Fundamental Issues in Computer Science*.

Luciano Floridi is professor of philosophy at the University of Hertford-shire (Research Chair in Philosophy of Information and UNESCO Chair in Information and Computer Ethics) and fellow of St Cross College, University of Oxford. In 2010, he was elected fellow of the Center for Information Policy Research, University of Wisconsin. His most recent books are: *The Cambridge Handbook of Information and Computer Ethics* (CUP, 2010), *Information: A Very Short Introduction* (OUP, 2010) and *The Philosophy of Information* (OUP, 2011).

Vincent F. Hendricks is professor of formal philosophy at the University of Copenhagen and Columbia University. He is editor in chief of the journal *Synthese*, and among his many books are *The Agency* (forth-coming), *Mainstream and Formal Epistemology* (2006), *Thought2Talk* (2006), *Formal Philosophy* (2005), and *The Convergence of Scientific Knowledge* (2001).

Gualtiero Piccinini is associate professor of philosophy at the University of Missouri—St. Louis. He works in the philosophy of mind and related sciences. Recent publications include "First-Person Data, Publicity, and Self-Measurement" (Philosophers' Imprint, 2009), "The Mind as Neural Software? Understanding Functionalism, Computationalism, and Com-putational Functionalism" (*Philosophy and Phenomenological Research*, 2010), and "Information Processing, Computation, and Cognition" (with Andrea Scarantino, *Journal of Biological Physics*, 2010).

Sherrilyn Roush is associate professor of philosophy at the University of California, Berkeley. She is the author of *Tracking Truth: Knowledge, Evidence, and Science* (Oxford, 2005) and more recently of "Randomized Controlled Trials and the Flow of Information," "Second-Guessing: A Self-Help Manual," "Closure on Skepticism," and "Optimism About the

Pessimistic Induction." Her main current project is on fallibility, self-doubt, and justified belief, which has led to a second-order generalization of Bayesianism that allows us coherently to acknowledge and adjust for information about our unreliability.

Andrea Scarantino is assistant professor of philosophy at Georgia State University. He works in the philosophy of mind, with primary focus on emotions. Recent publications include "Insights and Blindspots of the Cognitivist Theory of Emotions" (*British Journal for the Philosophy of Science*), "Core Affect and Natural Affective Kinds" (*Philosophy of Science*), and "Computation vs. Information Processing: Why Their Difference Matters to Cognitive Science" (with Gualtiero Piccinini, *Studies in History and Philosophy of Science*).

Gary M. Shute is associate professor of computer science at the University of Minnesota—Duluth. He has a B.A. in mathematics from South Dakota State University and a Ph.D. in mathematics from Michigan State University. He has taught a wide variety of computer science courses, including machine organization, computer architecture, operating systems, computer networks, data structures, software engineering, and object-oriented design. More recently, he has focused on understanding the human mind and its ability to make sense of the world.

Richard Volkman is professor of philosophy at Southern Connecticut State University and associate director of the Research Center on Computing and Society. His research evaluates the impact of information technologies on our abilities to lead the good life. Since the relevant information is decentralized, tacit, and local, this project involves articulating individualist moral and political philosophy for the information age and addressing the associated issues, such as intellectual property, identity, privacy, and digital culture.

1

PUTTING INFORMATION FIRST:
LUCIANO FLORIDI AND THE PHILOSOPHY OF INFORMATION

PATRICK ALLO

This collection of chapters is devoted to Luciano Floridi's contributions to the philosophy of information. As Floridi explains in his replies near the end of the collection, his own work during the past ten years has been almost co-extensive with *the* philosophy of information. Some might find this claim problematic. After all, doesn't the philosophy of information go all the way back to Leibniz? Or shouldn't we acknowledge the crucial role played by Wiener, Turing, Simon, Dretske, and many other fathers of the philosophy of information? Of course, the informational turn in philosophy cannot be reduced to work done in the past decade, let alone work done by one person (see the epilogue by Terrell Ward Bynum and the concluding section of Floridi's replies). Yet, what I want to emphasise is that while there has been a clearly discernible informational turn in recent (and not so recent) philosophy, Floridi gave that turn a more radical twist by claiming that taking the informational turn means redefining philosophy. This is why there is a conception of the philosophy of information that Floridi may rightly call his own.

In this introduction, I want to do two things: situate the contributions to this collection in the broader context of the philosophy of information, and say something more general about Floridi's unique understanding of what *the* philosophy of information is (or should be). I'll start with the latter, and then use my sketch of the philosophy of information as a map to situate the various chapters.

Let's start with an analogy. The philosophy of information is much like the philosophy of probability. The two have similar subject matter, a mathematically well-understood notion, but are stuck with widely divergent and often incompatible interpretations: chances, frequencies, subjective probabilities, propensities, and so on, for the former; qualitative, quantitative, syntactical, semantical, and so on, for the latter. If there is a core notion of information, then that core is composed only of formal properties, with Shannon's communication theory doing the unifying work. Likewise, classical probabilities satisfy the Kolmogorov axioms, but these do not settle what probabilities are. In both cases, the unification doesn't immediately extend beyond the logico-mathematical

framework. We can expand this analogy even further. The philosophy of
probability isn't just about interpretations but also covers philosophical
applications of probability theory like Bayesian epistemology and philosophical issues that arise from the application of probability theory.
Similarly, the philosophy of information isn't exclusively about the
concepts of information and computation, it is also concerned with
applications of informational and computational methods and models
to philosophical problems, as exemplified in this definition by Floridi:

> [T]he philosophical field concerned with (a) the critical investigation of the
> conceptual nature and basic principles of information, including its dynamics,
> utilisation, and sciences, and (b) the elaboration and application of informa-
> tion-theoretic and computational methodologies to philosophical problems.
> (Floridi 2002, 137)

Such applications include, for example, the formulation of an information-
based epistemology (Floridi 2006 and forthcoming), the use of the notion
of *strongly semantic information* in an analysis of relevance (Floridi 2008a),
and the defence of an informational ontology as a means to resolve a
central debate between two brands of structural realism (Floridi 2008b).

So far, the analogy seems quite compelling. When compared to the
picture of the philosophy of information that arises from a recent
handbook (van Benthem and Adriaans 2008), the analogy is in fact quite
accurate. This is because the philosophy of information is most commonly understood as referring either to a number of philosophies of
information or to the philosophical study of different formal theories of
information. When Floridi refers to *the* philosophy of information, he has
something quite different in mind, something that is at once more specific,
larger in scope, and more ambitious. In short, he thinks that there is just
one philosophy of information, that it doesn't apply merely to the study
of information in the narrow sense (i.e., the technical notion), and that its
ultimate goal is to transform the nature of philosophical theorising.
Clearly, the philosophy of probability doesn't have any of these features.
This is why the analogy is instructive. It illustrates where Floridi's use of
the label "philosophy of information" diverges from the generic use of
that label. Unless otherwise stated, I use the term "philosophy of
information" (PI) in this introduction in the sense intended by Floridi.

Let me illustrate the specificity of PI by means of two common
misconceptions about it. Misconceptions about PI often go hand in
hand with the objection that there is no such thing as *the* philosophy of
information. This doesn't mean that all these misconceptions are actually
found in print. Though some are inspired by actual opinions, they are
better thought of as typical instances of a more general attitude. In most
cases, the relevant objections are right about specific features of PI but
misconstrue or underestimate the range of the whole enterprise.

Correcting the received view is one thing, but for purposes of situating the different chapters in this collection I need a more positive view of PI. To that end, I describe it along three dimensions: (a) its subject matter, (b) its methods, and (c) the substantial views it propounds. This is already implicit in Floridi's claim that PI is a mature discipline because it introduces unique topics and original methodologies, and leads to new theories (Floridi 2002, 124).

As we shall see, the chapters in this collection engage PI along one or more of these dimensions. But first let us turn to the misconceptions.

First Misconception

Because there is no agreement on a unified notion of information, there cannot yet be an agreement on what *the* philosophy of information is. What this objection presupposes is that there is only one reading of "the philosophy of information," namely, as the philosophy of the one true concept of information. Hence, as long as this unique concept is missing there cannot be a single philosophy of information either.

The real misconception in this case isn't just the disputable assumption that a unified theory of information is desirable or philosophically fruitful, it is the stronger claim that the unification of a field or discipline like PI should always and exclusively follow from the unification of its subject matter, namely, the concept of information itself. This is highly problematic. Because of an all too narrow identification of PI with its core subject matter, it can only be one of two things: an exceedingly reductive enterprise or an eclectic collection of philosophies of information. Such pessimism is not justified. With regard to its subject matter, Floridi emphasises integration instead of unification: "On the whole, its task is to develop not a unified theory of information but rather an integrated family of theories that analyse, evaluate, and explain the various principles and concepts of information, their dynamics and utilisation, with special attention to systemic issues arising from different contexts of application and interconnections with other key concepts in philosophy, such as being, knowledge, truth, life, and meaning" (Floridi 2002, 137). More important, this quote also emphasises that the search for integration should not exclusively be grounded in the nature of information but should also reflect the uses of the information. That is, it should not only focus on what information really is but also shed light on why information is a cognitively valuable commodity and how the concept of information can play an explanatory role in the sciences. Integration, then, is about showing how different notions of information can play different and often complementary roles, both in our own epistemic and intellectual lives and for the working scientist. Obviously, such integration no longer requires a unified notion of information.

Second Misconception

The informational turn in philosophy is in the first place a trend that cuts across several (often technically oriented) sub-disciplines, especially epistemology and the philosophy of mind and language. So even if we can have an integrated conception of information, this integration will not diminish the fact that there is no single discipline called the philosophy of information. Since the integration is driven by methodological concerns (the use of information as a commodity and as an explanatory notion), PI can only play an auxiliary role. At its best, it is something that will be integrated into normal philosophical theorising. At its worst, it is a passing fashion with no enduring influence.

Whether or not PI will have a lasting influence is something only the future will tell. Yet, if it has any influence, it shouldn't be reduced to mere methodological recommendations. This is because the methodology of PI comes with more substantial philosophical insights, still methodological in nature but not neutral on substantial issues like the nature and purpose of philosophical theorising. The methodological canon of PI-style philosophy—the *method of abstraction, minimalism*, and *constructionism*—is indeed packed with views as to which questions are worth pursuing, and how they should be pursued. Since it would take us too far to show how each of these methods favours particular types of questions and recommends a specific way of answering them, let me make use of an example. Consider the following quote from Popper's "Epistemology Without a Knowing Subject": "An objectivist epistemology which studies the third world [the objective content of thoughts] can help to throw an immense amount of light upon the second world of subjective consciousness ... ; but the converse is not true" (Popper 1968, 338).

Given that similar considerations inspire Floridi's own take on epistemology, we can use this as a proxy for his views. The question we should then ask is where the methodological recommendations end, and where the substantial views about the nature of knowledge begin. This isn't obvious. Clearly, saying that our theory of knowledge shouldn't be clouded by mentalistic considerations is a methodological consideration; it can be considered as an application of *minimalism* to the problem of knowledge. Yet, since it immediately rules out most traditional theories of knowledge, the philosophical import of minimalism exceeds what can be described as mere methodology.

The lesson we learn from this second misconception is that a characterisation of PI in terms of its subject matter and methodological recommendations isn't entirely adequate. The problem, as we have seen, is that because it underestimates the philosophical import of the methodology of PI, such characterisation cannot account for the radical change PI stands for. With the introduction of new methods and concepts, the nature of philosophy changes as well.

This suggests a three-dimensional characterisation of PI, based on subject matter, method, and substantial views about the aims and purposes of philosophical theorising. Whether or not a substantial view defines the nature of PI is a critical issue. Clearly, not every view defended by Floridi should be situated at the core of PI. In my view, the demand to provide answers rather than analyses (see Floridi's replies) belongs at the core. By contrast, the much-disputed veridicality thesis (Floridi 2004 and 2005b) belongs to the periphery. It is a crucial part of Floridi's own work, but not something that defines the kind of theories PI should favour. Somewhere in between, then, we find opinions about what a theory of knowledge should be like, or rather what it shouldn't be like. I'd like to situate this fairly close to the core, but for most purposes it doesn't really matter where we draw the line. The point is rather that whatever belongs to the core, and this includes substantial views, shouldn't be left out of the characterisation of PI.

The Chapters

With the three dimensions of PI in mind, the various chapters in this collection can now be situated relative to how they engage with PI. Some focus on specific topics within it, others adhere to its methodology or specifically challenge views previously defended by Floridi. But most engage with PI along multiple dimensions at the same time. This is the most fruitful approach.

The two chapters in the section on knowledge, by Roush and Hendricks, respectively, do not really challenge Floridi's work as such but make independent contributions to PI. This is primarily because they adhere to methods that are "minimalist" without being simplistic. Hendricks does so within his own distinctive approach of modal-operator epistemology, and presents the problem of pluralistic ignorance within that setting. The distinctive feature of his approach is the view that knowing doesn't merely require information; it also requires the ability to process it. Roush, by contrast, uses the idea of the knowledge game (Floridi 2005a) to show how knowledge can be more valuable than mere true belief. She combines Lewis's signalling games with considerations about evolutionary stable strategies to show that at least a tracking account of knowledge does not face the so-called *swamping problem*.

Bringsjord picks up the theme of the knowledge game as well, but sticks to its original purpose: showing how one can know that one is not a zombie. By taking the idea that the question "How do you know you're not a zombie?" is best answered in terms of passing a test, he implicitly subscribes to the methodological assumptions on which Floridi has relied. By being more optimistic about the prospects of artificial intelligence (AI), with a proof to support the claim that logic-based AI could pass

Floridi's most elaborate test, Bringsjord also challenges Floridi's own views.

The next two chapters engage PI at its traditional core: the concept of information and its relation to knowledge. Scarantino and Piccinini take a new look at what is perhaps Floridi's most debated view: the suggestion that semantic information is by definition truthful, a view previously defended by Dretske and Grice. Dretske warns that "the tendency in computer science to construe information as anything—whether true or false—capable of being stored on a hard disk is to confuse non-natural meaning (or perhaps just structure) with genuine information. It leaves it a mystery why information should be thought a useful commodity" (Dretske 2009, 383). Scarantino and Piccinini, however, take the distinction between natural and non-natural meaning precisely as their starting point, and argue for a more refined view of semantic information, one that is closer to what cognitive scientists have in mind. Adams initially stays close to Floridi's analysis of information, and its place within the informational turn in philosophy, by comparing his analysis of "being informed" with the distinctive rejection of closure we find in Dretske's theory. That's only one part of Adams's chapter, which also engages the *symbol-grounding problem* and questions the stringent constraints Taddeo and Floridi 2005 imposed on any solution to that problem.

Colburn and Shute work in the philosophy of computer science, and in their contribution to the collection they use the notion of a law, as it is used by computer scientists (i.e., as invariant), to investigate two aspects of Floridi's work. The first is purely methodological, namely, the method of abstraction; the second is more substantial, namely, the notion of a purely computational or informational reality. In both cases, Colburn and Shute reveal close connections and draw our attention to points at which Floridi's use of these notions diverges from their mainstream use in computer science.

Bueno's contribution also deals with the idea of an informational reality, but approaches it from the perspective of structural realism (as it is presented in Floridi 2008b). The main question Bueno asks has to do with how Floridi's proposal leads to a realist position. His answer is based on an alternative proposal: using partial structures and the notion of quasi-truth. The upshot of the proposal is that ontic and epistemic structuralism can still be reconciled, but that an empiricist reading becomes available as well.

Finally, Volkman discusses an aspect of Floridi's work that has already been a topic in a collection on Floridi (Ess 2008) but could hardly be omitted here: information ethics. In Volkman's view, the valuable part of information ethics is its emphasis on *good construction*, and the disputable part is the value it accords to impartiality and the emphasis on foundations. This is why, according to Volkman, information ethics may learn some valuable lessons from virtue ethics.

Conclusion

By way of conclusion, I'd like to do two things. First, recommend the final chapter, by Terrell Ward Bynum, which serves as an epilogue for this collection and is the perfect complement to the specific points I've tried to make in this introduction. And second, emphasise once more what I think is the real challenge for the philosophy of information: namely, to have a concrete and lasting influence by engaging the philosophical community at large, and to avoid either becoming a narrow subfield—a philosophy of the concept of information—or ending up as a set of methodological recommendations devoid of real philosophical substance.

Acknowledgments

I would like to thank everyone who contributed to the successful completion of this project, including all the contributors and referees and the editorial team of *Metaphilosophy*, and Wiley-Blackwell. Special thanks go to Otto Bohlmann for his patience and logistical support during the final stages.

References

Dretske, Fred. 2009. "Information-Theoretic Semantics." In *The Oxford Handbook of the Philosophy of Mind*, edited by Brian McLaughlin, Ansgar Beckermann, and Sven Walter, 318–93. Oxford: Oxford University Press.

Ess, Charles, ed. 2008. *Luciano Floridi's Philosophy of Information and Information Ethics: Critical Reflections and the State of the Art*, special issue of *Ethics and Information Technology* 10, no. 2.

Floridi, Luciano. 2002. "What Is the Philosophy of Information?" *Metaphilosophy* 33, nos. 1–2:123–45.

———. 2004. "Outline of a Theory of Strongly Semantic Information." *Minds and Machines* 14, no. 2:197–222.

———. 2005a. "Consciousness, Agents and the Knowledge Game." *Minds and Machines* 15, nos. 3–4:415–44.

———. 2005b. "Is Information Meaningful Data?" *Philosophy and Phenomenological Research* 70, no. 2:351–70.

———. 2006. "The Logic of 'Being Informed'." *Logique & Analyse* 49, no. 196:433–60.

———. 2008a. "Understanding Epistemic Relevance." *Erkenntnis* 69, no. 1:69–92.

———. 2008b. "A Defense of Informational Structural Realism." *Synthese* 161, no. 2:219–53.

———. Forthcoming. "Semantic Information and the Network Theory of Account." *Synthese*.

Popper, Karl R. 1968. "Epistemology Without a Knowing Subject." In *Logic, Methodology and Philosophy of Science III*, edited by B. van Rotselaar and J. F. Staal, 333–73. Amsterdam: North-Holland.

Taddeo, Mariarosaria, and Luciano Floridi. 2005. "Solving the Symbol-Grounding Problem: A Critical Review of Fifteen Years of Research." *Journal of Experimental and Theoretical Artificial Intelligence* 17, no. 4:419–45.

van Benthem, Johan, and Pieter Adriaans. 2008. *Handbook on the Philosophy of Information*. Amsterdam: Elsevier.

2

THE VALUE OF KNOWLEDGE AND THE PURSUIT OF SURVIVAL

SHERRILYN ROUSH

In honor of Luciano Floridi, for his leadership in bringing information theory to epistemology

It is thought that externalist views of what knowledge is—that do not require conscious access to reasons and arguments but only certain relations in which a person's belief must stand to the world—have trouble explaining why knowledge is more valuable than mere true belief. Justificatory arguments are intrinsically valuable, some say, but what additional epistemic worth could another relation of your belief to the external world have if the belief already has the relation of being true? (Swinburne 1999, 2000, Kvanvig 2003)

This surely depends on what the further relation is. The value problem, or "swamping problem" as it is called because the property of truth of the belief seems to swamp in significance other external relations, appears particularly acute for process reliabilism, but process reliabilism is not the only externalist view. Besides truth the process reliabilist puts constraints only on the history of formation of a belief, but once you actually *have* a true belief the extra property of its having been formed in a certain way seems otiose. According to the standard comparison, a beautiful chair does not have additional aesthetic worth for having been produced by a process that produces beautiful chairs most of the time.[1] If these points stand, then the swamping problem appears to afflict all historical views of knowledge.

Not all externalist views of knowledge are historical, however. Counterfactual views do not impose conditions on the genesis of a belief. They put conditions on how the belief is currently disposed to behave or fare in scenarios different from the actual one. The distinction here is analogous to the difference between how the solar system was formed and what laws govern its motions. The laws govern the motion of a planet at every point in time, even if the planet is not in fact moving. The history of its

---

[1] This seems right, but it is curious that beautiful chairs having been designed or made by an artist or establishment who usually makes very beautiful chairs tends to make the chair sold under that label have higher market value.

formation is a different matter; it will conform with these laws but involve a lot more information, about initial conditions, for example. This is where the analogy ends, of course, for we, unlike the law-governed physical world, are capable of forming beliefs by processes that are not well behaved in the relevant way, and of forming beliefs that do not conform in their dispositions to any epistemologically nice counter-factual properties. The point is that these are distinct failures. Even if there are correlations between them in the actual world, as there probably are, those would be contingent relationships. The first failure is a defect of the process of forming a belief, the second a defect in the product. The process reliabilist thinks that the first type of failure—formation by a process that does not tend to produce true beliefs—is what deprives a belief of the status of knowledge even if it happens to be true. A counterfactualist thinks the failure of a true belief to be knowledge is a defect in the dispositions that accompany the fully formed belief.

Since ascription of a counterfactual concerning a belief is ascription of a current property, counterfactual views of knowledge are "current time-slice" views, in Alvin Goldman's terminology (Goldman 1979), a property they share with traditional internalist justified-belief views. Accordingly, the value problem for these views looks entirely different from that for process reliabilism. Here the question becomes whether a person's disposition to believe or not believe a proposition p in nonactual situations could add value when she already actually has a true belief in p. It is clear intuitively that counterfactuals might have something to offer here. After all, your spouse's not actually having an affair with Mr. or Ms. X is a good thing, but it would surely be strictly better if it were also the case that he or she wouldn't have that affair even if offered a million dollars. The latter is evidently not swamped by the former.

Love might be a case where the counterfactual enhances an intrinsic kind of value. I will argue that if knowledge requires tracking then it enhances the extrinsic value of a true belief, the value it has for achieving or obtaining other things. That is, it will turn out that Socrates was wrong to think that knowledge of which road went to Larissa would be no more valuable than mere true belief about it. I will show that the additional dispositional properties required by the tracking view of knowledge, formulated using conditional probabilities rather than counterfactuals, add payoff and survival value necessarily and that no other conditions on knowledge have the property that ensures this necessarily. This follows because tracking is the unique Nash Equilibrium and Evolutionarily Stable Strategy in what I will call the True Belief Game.

Intuitively, what fulfillment of the tracking conditions adds to the truth of a belief is a kind of robustness against contingencies. The main question in taking this analysis as a resolution of the value problem is whether robustness of a person's belief behavior as the subject is faced with a world that evolves over time is of value to a person at the time of holding the belief. I will discuss this after explaining the kind of robustness in question.

To begin, consider the situation of a person playing a game with the world, which I will call Nature, on a single occasion. Nature can play p or $-p$, p a proposition, and the person can play $B(p)$, that is, believe p, or play $-B(p)$, that is, not believe p. Suppose the person's payoffs are positive if he plays $B(p)$ when Nature plays p, and positive when he plays $-B(p)$ to Nature's $-p$, and they are negative when he plays $-B(p)$ to Nature's p and when he plays $B(p)$ to Nature's $-p$. These payoffs express the conditions that when p is true, it is more valuable to the subject to believe p than not to believe p, and when p is false it is more valuable to the subject to not believe p than to believe p. The results I am explaining are limited to these conditions, but that is not a limitation on their application to the value problem. There are plenty of p for which it is more valuable to have a false belief than a true belief or no belief—think of crazy metaphysical beliefs that come bundled with other, true, beliefs holding all of which is required to cement your relation to your social group, and consider a situation (of pioneers, for example) where survival depends on membership in a group. However, these cases of p for which true beliefs are not valuable are not relevant to the value problem under discussion here, which is to say whether or how given that true belief is valuable, knowledge has added value. Anyway, cases of p for which true beliefs are not valuable are not cases where we would expect knowledge to be valuable either.

Nature is indifferent to what you play when it plays p or $-p$. It gains nothing and loses nothing, so in our first pass here we are dealing with a degenerate game:

	$B(p)$	$-B(p)$
p	$(0, 2)$	$(0, -3)$
$-p$	$(0, -1)$	$(0, 3)$

Nature is the player choosing a row, and the subject is the player choosing a column. The winning strategy for the subject is to play $B(p)$ when Nature plays p and to play $-B(p)$ when Nature plays $-p$, which is reflected in the payoffs in the four possible situations written as ordered pairs with the subject's payoffs second.[2] The word "strategy" does not

---

[2] The particular numbers are important here only for some of the ordinal relationships: the payoff for a true belief must be greater than that for no belief when p is true, and no belief

imply, in this or any other game I discuss here, conscious or consciously accessible planning, or even thought. It is simply an intuitive term for what the player does; later in the discussion strategies will be rules in accord with which players act in a given round of play, and we will discuss dispositions to act in accord with a given rule, but doing or having any of these does not require conscious access to them either. We do not assume that the subject knows (or does not know) what Nature played before playing $B(p)$ or $-B(p)$.[3] Nor are the players assumed to know the structure of the game or their or their opponent's payoff structure. We think merely of which of their options the players play, and what they get when they do. This game can thus be compared with Floridi's more sophisticated Knowledge Game (Floridi 2005). Here we will see how knowledge emerges from a game in which the object is true belief and no knowledge of any sort is assumed. There common knowledge is assumed in order to show how second-order knowledge that one is conscious can emerge from the fact that there is a game one can win that a zombie could not, and that one can see that one can win it and a zombie could not.

This simple game is a way of formulating what is essential in forming beliefs about matters like p in situations where the truth of p matters positively to us; it formulates the starting point of the value problem. The game expresses the assumption that at a given time a true belief about p is not trivially acquired and is valuable by saying that whether p is true or false is determined not by the subject but by a different player, Nature, and that a correct belief state about p (whether that means believing it or not believing it) has positive value for the subject; it makes her win the game and achieve her best possible outcome given Nature's play.

Signaling Games and Repeated Play

We have seen that the assumption that a merely true belief has value can be represented as a certain kind of payoff structure in a one-shot game played with Nature. In real life, even with a single p, this game would often be repeated over time. The truth value of p may change; the tiger may be gone today but back tomorrow. We will imagine this repeated play with p and $-p$ understood as states of the world, and belief and absence of belief in p understood as acts of the subject. I will represent this as a more elaborate type of game, which I will introduce using an example

must be of greater value than belief when p is false. How much greater may be different with the two types of mistake one might make, depending on how costly a false positive or false negative is for the subject.

[3] Such an assumption would anyway trivialize the representation. The question would then be: If the subject knows that Nature made p true, should she believe p? Since knowledge implies true belief, the answer would be automatic— she already would believe it—and the representation would swing independent of whether true belief has value or not.

about something other than belief. In this game, the states of the world may be different at each round of play, and the players may opt for different acts at each round of play. If we also imagine messages interposed between states of the world and acts of the subject, then we have what David Lewis called a "Signaling Game" (Lewis 1969). Such a game has the following kind of structure:

States of world	Messages sent	Acts
p	m_1	Watching football
q	m_2	Self-reflection

There is a Sender and a Receiver in the game, and each will have payoffs associated with strategies for responding to scenarios. Sender is defined by her repertoire of possible plays: here either m_1 or m_2 when Nature sets p or q. Receiver is defined by his repertoire of possible acts, here Watching football or Self-reflection, when Sender plays m_1 or m_2. In order for there to be a game of this sort, Sender and Receiver must each have the capacity for a variety of rules of responding when the other player plays in each of the possible ways he or she might. Thus, if the possible messages are m_1 and m_2, Sender must be able to respond by taking p to one of these, and taking q to one of these. Sender may send both states to one message or send p and q to different messages and it still be a Signaling Game. Similarly, Receiver must have the ability to act either of his two ways, Watching football or Self-reflection, and in any of the four possible permutations of rules for responding to each of m_1 and m_2.

A full set of such rules, that is, one that covers the possibilities a player may be paired with, is called a *strategy*. Thus Sender's strategy is a set of two rules, for example, T_1:

$$p \rightarrow m_2$$
$$q \rightarrow m_1$$

Likewise for Receiver, for example, L_1:

$$m_1 \rightarrow \text{Watch football}$$
$$m_2 \rightarrow \text{Watch football}$$

In a picture:

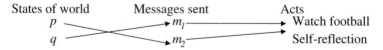

States of world   Messages sent   Acts
    p m_1 Watch football
    q m_2 Self-reflection

Sender and Receiver each have a repertoire of other possible strategies that is easy to list:

T_2:	$p \rightarrow m_1$	$q \rightarrow m_1$
T_3:	$p \rightarrow m_2$	$q \rightarrow m_2$
T_4:	$p \rightarrow m_1$	$q \rightarrow m_2$
L_2:	$m_1 \rightarrow$ Self-reflection	$m_2 \rightarrow$ Self-reflection
L_3:	$m_1 \rightarrow$ Watch football	$m_2 \rightarrow$ Self-reflection
L_4:	$m_1 \rightarrow$ Self-reflection	$m_2 \rightarrow$ Watch football

We can assess all possible outcomes for Sender and Receiver in this game by looking at the payoffs for their possible strategies when each is paired with each possible strategy of the other player. This is because a strategy pair, one from each player, determines what each will do whether the world is p or the world is q, the only two world states stipulated to be relevant to our game.

Thus, we consider payoffs for all combinations of T_1, T_2, T_3, T_4 with L_1, L_2, L_3, L_4. If we were modeling an actual situation, we would take these payoffs from the facts. How nice or nasty is a certain consequence for a given player, and how likely is that consequence if she plays a particular way and the other player plays a particular way? In this example I make the payoffs up, to illustrate some key points.

	L_1	L_2	L_3	L_4
T_1	$(-1, 2)$	$(2, 0)$	$(-2, -1)$	$(2, 0)$
T_2	$(-2, -2)$	$(0, 3)$	$(-1, 2)$	$(2, -1)$
T_3	$(0, 1)$	$(2, -1)$	$(-1, 2)$	$(1, -2)$
T_4	$(-2, -1)$	$(2, -1)$	$(3, -2)$	$(-2, 1)$

The first number of each ordered pair is the payoff for Sender when she plays the strategy to the left, and the second is what Receiver gets when he plays the strategy at the top of that row. The numbers could be anything, but the ones I have entered for the current example imply that, for example, if Sender plays T_3 and Receiver plays L_2 then Sender gains 2 and Receiver loses 1. If Receiver plays L_4 to Sender's T_4, Receiver gains 1 and Sender loses 2. A feature that I have written into this particular assignment of payoffs is that there is no one combination of plays (square in the table) that will have both Sender and Receiver better off compared to their other options in that row or column, respectively. This implies that if we set Sender and Receiver to play the game in perpetuity, neither of them would settle on one strategy out of their repertoire.[4]

Although the definition of these games does not involve any assumption of intentionality, knowledge, or information transmission, the terms "message," "sender," and "receiver" are meant to be suggestive. This is because phenomena we recognize as information transmission can arise naturally out of the games. We can see how by looking at what the scenario just imagined means intuitively. Suppose Sender plays T_1 and Receiver L_1, as in the picture above. Sender's dispositions in T_1 mean she is cued in to there being a difference between p and q and is revealing that difference by differentiating between them in a uniform way in the messages she sends out. Receiver's dispositions in L_1 mean he does not register that difference in his act of Watching football or Self-reflection. Receiver's indifference to the distinction between m_1 and m_2 in his responding act makes the information about the state of the world, p or q, unavailable to Receiver; we could say that he is not listening. According to our payoff table, and assuming a simple dynamics, it follows that this set of dispositions is not a configuration our two players would stably end up in, since though it is beneficial to Receiver not to hear ($+2$), it is detrimental for Sender not to be heard (-1), and this is not better than all the players' other options. We can imagine this as a realistic case where Sender has an interest in communicating and Receiver has an interest in not hearing. Since every other square also involves some analogous mismatch of best interest, it follows that, other things equal, this relationship will not become stable no matter how many rounds they play.

The basic condition under which they would become stable is there being a square where both receive the highest payoff they could get given the strategy the other has played. This would happen here, for example, if

---

[4] That is, since there isn't a strict Nash Equilibrium there will be no evolutionarily stable strategy (ESS). The notion of ESS makes sense for this asymmetric game if we think of ourselves as referring to its symmetric counterpart in which every player has both a Sender strategy and a Receiver strategy that he or she plays depending on whether he or she is assigned the role of Sender or Receiver in a given round of play. All of the claims about stability in the Watching football game should be taken to be referring to that symmetric game.

the top left corner had the payoffs (4, 4). If in the course of play the players happen on such a combination, and stick with it for a while, then they will become stable and resiliently wedded to the corresponding strategies indefinitely.

That there be a possibility of stable convergence depends on there being a payoff possibility reflecting common interest. However, there being a common interest does not imply that the interest is in what we would intuitively call communication, or information transmission. Witness that nothing prevented us imagining the highest mutual payoff in the square T_1, L_1 where Sender makes information available but in fact is never heard. A strategy combination where Sender talks into the wind can become stable as long as Sender is satisfactorily rewarded in it. Whether a stabilizable configuration is also communicative depends on which strategies in fact have the highest payoffs. Thus, the issue of stability and the issue of communication are conceptually independent. The reason the new payoff structure leads to convergence to a single set of strategies is that if the top left corner is (4, 4), then this option dominates every other possible set of strategies. That is, it is better, for both players, than any other option either of them has. This dominating ordered pair of strategies is called a strict Nash Equilibrium, and such a pair of strategies in a Signaling Game is called a Signaling System. As we have just seen, not every Signaling System will lead to what we would intuitively call communication or information transmission.

The strategy pairs in our game that would intuitively correspond to information transmission are $<T_1, L_3>$, $<T_1, L_4>$, $<T_4, L_3>$, $<T_4, L_4>$. In all of them Sender can tell, and signals, the difference between p and q, and Receiver can distinguish those messages and respond differentially. Notice that the messages m_1, m_2 have no preassigned meanings. As long as Sender consistently sends the same message for p each time and q each time, and sends different messages for p and q, it does not matter which of m_1 or m_2 she uses in which role. As regards whether it is possible to achieve a Signaling System or information transfer the qualities of m_1 and m_2 are conventional. Receiver will be able to cotton on to p versus q if he has available a strategy that can respond differentially to m_1 and m_2.

Responding *differently* to p and q is one thing, one might say, but what about responding correctly? If the character of the messages is completely conventional, couldn't Receiver get his responses to p and q exactly the wrong way around? The answer is no. The *right* response to p for Receiver is determined by his payoff structure—how good it is for him when he watches football in the situation p. But his payoff structure also determines which combined sets of strategies of him and his playmate can become stable Signaling Systems. Thus, here are the four possible intuitively communicative Signaling Systems in our game:

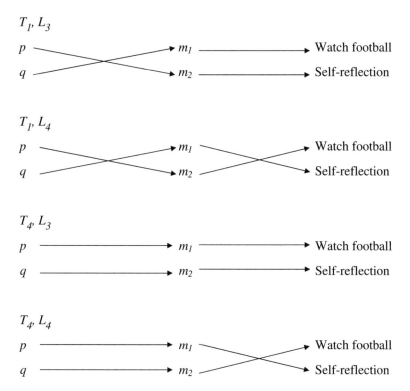

Suppose it is appropriate for Receiver to respond to state p by Watching football and to q by Self-reflection. These assumptions will be reflected in his payoffs. His rewards will be highest, among all the possibilities, for $<T_1, L_4>$ and $<T_4, L_3>$. The rewards will be the same for both of those strategies because each has him successfully responding to p with Watching football and to q with Self-reflection; it doesn't matter which message route he took to get there. Accordingly, if all of these communicative solutions are winners for Sender, then, according to theorems, Sender and Receiver will be stable in either of $<T_1, L_4>$ and $<T_4, L_3>$. Receiver will have a system in which he is able to make what is for him the appropriate response to p, because the appropriateness will be reflected in his payoffs, and which system is stable, if any, is determined by the payoffs for him and the other player.

The True Belief Game as a Signaling Game

To model the task of getting true beliefs about the physical world as a Signaling Game with two players and repeated play, we take the possible

States of the World to be p and $-p$. We take Sender to be the laws of Nature, those things that determine which indicators flow downstream from the State of the World, p or $-p$. Receiver is a player about whom we stipulate the values of true belief, false belief, and no belief as in the value problem about knowledge, thus, as in the very first table above. Sender and Receiver each have four possible strategies:

Sender:

N_1:	$p \rightarrow M_1$	$-p \rightarrow M_2$
N_2:	$p \rightarrow M_1$	$-p \rightarrow M_1$
N_3:	$p \rightarrow M_2$	$-p \rightarrow M_1$
N_4:	$p \rightarrow M_2$	$-p \rightarrow M_2$

Receiver:

K_1:	$M_1 \rightarrow B(p)$	$M_2 \rightarrow -B(p)$
K_2:	$M_1 \rightarrow -B(p)$	$M_2 \rightarrow -B(p)$
K_3:	$M_1 \rightarrow B(p)$	$M_2 \rightarrow B(p)$
K_4:	$M_1 \rightarrow -B(p)$	$M_2 \rightarrow B(p)$

	K_1	K_2	K_3	K_4
N_1	$(0, 4)$	$(0, 1)$	$(0, -2)$	$(0, -5)$
N_2	$(0, -1)$	$(0, -1)$	$(0, -1)$	$(0, -1)$
N_3	$(0, -5)$	$(0, 1)$	$(0, 1)$	$(0, 4)$
N_4	$(0, -1)$	$(0, -1)$	$(0, -1)$	$(0, -1)$

The payoffs in this table are assigned by thinking of values not as possessed by the strategies but as possessed by particular types of outcomes, in our case the reward or punishment for the having or lacking of a true belief. The payoffs could not belong to the strategies per se, because those achieve different payoffs depending on the strategy played by the other player. We assume that a merely true, particular belief p occurring at a particular time is valuable, more valuable than no belief when p is true, and so on, and payoffs in our tables always refer to that level of fact. A strategy, which I will eventually associate with that part of knowledge that goes beyond true belief, is obviously instrumentally valuable insofar as it actually gives you one of these valuable things. However, that fact alone would not resolve the swamping problem, since it would still be the case that if you had that true belief you wouldn't need that instrument. What I will eventually show is that a strategy, which in a distinctive way attaches to an actual true belief, adds value beyond that which comes from the truth of the particular belief it is a good tool for getting.

Assuming true belief, false belief, and no belief have the relative values we stipulated at the start, the worst combinations for our subject would be $<N_3, K_1>$ and $<N_1, K_4>$. Both combinations have Receiver believing p when it is false and not believing when p is true, and those belief states are bad for him. The best combinations for Receiver are $<N_1, K_1>$ and $<N_3, K_4>$, since in these cases Receiver believes p when it is true and does not believe p when it is false, and those outcomes are good for him. In our True Belief Game, Receiver's payoffs happen to be the worst and the best, respectively, in these two strategy sets.

Nash Equilibria and ESS in the True-Belief Signaling Game

Whether the repeated True-Belief Signaling Game converges to a stable Signaling System depends not only on Receiver's payoffs but also on the payoffs of the other player, Nature. It is commonplace in proving epistemological convergence theorems to assume that Nature is cooperative in making separating evidence available, that is, in providing distinct indicators of distinct states of affairs. We cannot expect a subject to gain information about Nature if she obfuscates. The question is then whether the subject's tools will enable him to find or use the messages appropriately. Here this assumption will correspond to Nature's having a preference to play strategy N_1 or N_3, indifferently between the two, regardless of anything else, since in those two strategies, and only those, distinct messages are given for distinct states of affairs. Writing that in:

	K_1	K_2	K_3	K_4
N_1	(1, 4)	(1, 1)	(1, −2)	(1, −5)
N_2	(−1, −1)	(−1, −1)	(−1, −1)	(−1, −1)
N_3	(1, −5)	(1, 1)	(1, 1)	(1, 4)
N_4	(−1, −1)	(−1, −1)	(−1, −1)	(−1, −1)

The two best outcomes for Receiver occur in blocks that are also best outcomes for Sender, $<N_1, K_1>$ and $<N_3, K_4>$, but Sender is indifferent between those. These squares are Nash Equilibria because neither player can do better, given that play of the other, by going to some other square. However, they are not strict, because neither of these dominates all options for Sender.

We can find a strict Nash Equilibrium by focusing more closely on the fact that in our game Sender is Nature. It is her laws that determine how she responds to states of the world and produces indicators for them; N_1 and N_3 are two different possible sets of laws that are equally capable of delivering separating evidence, and that is all we took to matter to her. Yet in a real game Nature, as usually conceived, does not change her laws with each round in a given domain of knowledge-seeking. She would have chosen N_1 or N_3 in the beginning, and the repeated game would be a degenerate one that Receiver plays against the background of that one strategy. (In a moment we will see that he plays it in competition with other receivers.) If we assume that the world has only one set of physical laws, then either N_1 or N_3 will be the unique play Nature always makes. If so, then either K_1 or K_4 (depending on which laws Nature chose) will be the unique best response of Receiver, and either $<N_1, K_1>$ or $<N_3, K_4>$ will be a strict Nash Equilibrium.

Since there is symmetry, we can suppose without loss of generality that Nature chose strategy N_1. The state of the world remains as it was, potentially changing in each round between p and $-p$. Now Sender (Nature) is a degenerate player who is like a set of background conditions, and the game that is left involves a confrontation between the Receiving strategies. We can rewrite this as a nondegenerate game by imagining many Receivers playing with each other. They meet two by two, round by round, each does his thing with Nature and each gets a certain payoff determined by what his payoff was in the previous table when playing against N_1. It is not that these two players necessarily interact or oppose each other, simply that each may do better or worse than or the same as the other with which he is paired in a given round; each competes to get higher payoffs than the opponent, as in darts, but not necessarily as in football. What we are now doing is comparing what one's outcomes in the True Belief Game would be were one to be this kind of Receiver or that. There are four types of Receiver, each defined by his strategy when faced with N_1:

	K_1	K_2	K_3	K_4
K_1	(4, 4)	(4, 1)	(4, −2)	(4, −5)
K_2	(1, 4)	(1, 1)	(1, −2)	(1, −5)
K_3	(−2, 4)	(−2, 1)	(−2, −2)	(−2, −5)
K_4	(−5, 4)	(−5, 1)	(−5, −2)	(−5, −5)

K_1 dominates all other possibilities—it does better than every other possibility in the payoffs—and so is a strict Nash Equilibrium. It is a consequence of this that if you were to always play K_1, that is, play it in every round of the game, then you would always do better than if you had played any other possible strategy. That is, it is not only that a true belief has value but also that there is a unique strategy that will deliver a correct belief state no matter what, in particular no matter whether p is true or false (and even if the strategy is offered a million dollars to do otherwise). This added general guarantee over several dimensions of possible variation is what will yield an answer to our value question about knowledge, as I will discuss after relating the present concept of strategies to theories of knowledge.

In this latest game we imagined an individual player being of a certain type, corresponding to a strategy, and him having a true or false, or lack of, belief in each round. I will explain below how what his strategy is can add value, in each round, to his having actually achieved a true belief. This is an advantage that accrues to an individual when he is of a favorable type, but the very same facts also guarantee a value added for a population of individuals of his type. The reason is that a strict Nash Equilibrium in a symmetric game like this one is an Evolutionarily Stable Strategy (ESS). This is a notion used to evaluate the fate of subpopulations of uniform types, here four subpopulations for the four types of Receiver for whom true belief about p is valuable. The proportions of these types in the population change with each round as any individual player's strategy in the next round will be the one determined to be the best by set rules of interaction dynamics applied to his and possibly others' outcome(s) in this round. The question, as with biological evolution, is how the proportions of the four types evolve with each generation or round of play.

The interesting implication of a strategy's being an ESS is that if it comes to be widespread in the population, it will be uninvadable by a mutant strategy; that is, no other single strategy that exists or arose in small numbers could drive this type to extinction.[5] This is a powerful property because it holds, when it does, no matter what the dynamics of interaction are as the game evolves from round to round (and there are a potentially infinite number of possible interaction dynamics). The basic upshot of this for our case is that if the K_1 strategy gets a good start it will be the unique type that rebuffs every competitor in the True Belief Game.

The Value of Knowledge

The fact that the K_1 strategy is strict Nash, an ESS, and a Signaling System is the key to answering the value problem for the tracking view of

---

[5] *Two* player types could successfully gang up on an ESS, so an ESS is uninvadable but not unbeatable. The mixed strategies discussed below are not equivalent to two strategies ganging up in the relevant way.

knowledge. We can see what K_1 has to do with probabilistic tracking, and knowledge, by looking more closely at what a strategy is and what the winning strategies in the True Belief Game look like. Assuming, as above, that Nature chose N_1 for her laws, K_1 is an ESS and strict Nash Equilibrium (sNE):

N_1:	$p \;\rightarrow\; M_1$	$-p \;\rightarrow\; M_2$
K_1:	$M_1 \;\rightarrow\; B(p)$	$M_2 \;\rightarrow\; -B(p)$

Since N_1 is simply assumed to be the case, the player who always uses K_1 has a relation to the world, that is, to p and $-p$, that we can think of as a result of the combined rules N_1 and K_1. The arrows in these diagrams are normally written in terms of conditional probabilities. So, in the simplest terms, a commitment to following the K_1 strategy on assumption of N_1 would be written

$Pr(M_1/p)$ = very high, and $Pr(B(p)/M_1)$ = very high   *
$Pr(M_2/-p)$ = very high, and $Pr(-B(p)/M_2)$ = very high   **

Notice the similarity of * and ** to the two probabilistic tracking conditions, respectively:[6]

$Pr(B(p)/p)$ is high   †
$Pr(-B(p)/-p)$ is high   ‡

The two sets of conditions cannot be unconditionally identified, because conditional probability is not transitive. However, under the following natural screening-off conditions

$Pr(B(p)/p.M_1) = Pr(B(p)/M_1)$ and
$Pr(-B(p)/-p.M_2) = Pr(-B(p)/M_2)$

* and ** imply † and ‡, respectively. (See Appendix below.)[7] That is, *if you are a faithful follower of the strict Nash Equilibrium or Evolutionarily Stable Strategy for a given p in the True Belief Game, then you fulfill the tracking conditions on knowledge for that p.* This means that no theory of knowledge that does not impose the tracking conditions implies that knowledge gives us an ESS or sNE. If following the rule of an ESS or sNE adds value to having a true belief, then it is a value that only tracking can give. Other conditions on knowledge may have other (nontracking)

---

[6] See Roush 2005 for a fuller discussion of these conditions.
[7] These screening-off conditions say intuitively that M_1, M_2 are the only messages about p that Receiver responds to, which is how our game has been written. This simple representation can model more complex cases if we think of a disjunction of circumstances as one message.

properties that make knowledge more valuable than mere true belief and thus address the swamping problem, but those properties must be strictly logically weaker than ESS or sNE, and they will not dominate in the True Belief Game, a representation that does seem a fair way of depicting what our task is in forming beliefs.

To see the other direction of relationship between the tracking conditions and the True Belief ESS and sNE conditions, we must consider that the tracking conditions are highly abstract, even more abstract than our imagined Signaling Game. Magic could be the truth maker of the conditional-probability conditions if magic existed. They involve no requirements that there exist a process of belief formation, or causal connection, the things we familiarly use to get to a knowledge state. How a subject manages to achieve fulfillment of the tracking conditions is not restricted by these conditions for what knowledge is. However, it happens to be a contingent fact about human beings that we can't fulfill the tracking conditions without intermediaries: causal processes, one event indicating another, one trait correlated with another, our eyes, our brains, having dispositions to respond differentially, testimony of witnesses, and so forth. The minimal description of what these intermediaries give to us that is sufficient to ensure tracking is indicators playing the role of messages, M_1 and M_2 in a Signaling System. Thus, what we can say is that if a *human being* fulfills the tracking conditions for a given p then *there are* M_1, M_2 such that she has an ESS and an sNE.

Having had your belief formed through a reliable process does not imply that your belief also has the tracking properties. Nor does the counterfactual property of safety (Roush 2005, 118–26). Being justified in your beliefs, or virtuous, also does not yield tracking. (Neither do the advocates of these conditions intend them to.) These alternatives to tracking are all nice properties, but they do not give you an ESS. However, one might be bothered about the fact that none of those conditions implies strategy K_2, K_3, or K_4 either. Having used a reliable process doesn't ensure that you would believe p if p were true, but it doesn't guarantee that if it were true you *wouldn't* believe it either. Are we really comparing actual theories of knowledge at all in the True Belief Game?

None of the other theories' conditions on knowledge implies any single one of the four pure strategies, yet they—and every other possible set of conditions—are taken account of in our game. Any set of conditions for knowledge that a subject fulfills will have consequences for whether or how often, or likely it is that, the subject will end up believing p when p is true and avoiding belief when p is false, with possibly different probabilities depending on a variety of conditions. Having formed your belief in p through a reliable process need not imply that you will believe p when it is false in order to confer some probability of doing so given the way the actual world works, or under certain conditions. Being justified in whatever way one prefers may not imply one's avoiding belief in p when p is false given the way the actual world works, but there may be some, even significant, probability, x, of it, and thus a $1 - x$ probability of believing p when p is false.[8] Similarly for the other rules in the True Belief Game. In this way, any conditions on knowledge that are added to the truth and belief conditions is represented as some "mixed strategy" in the True Belief Game. It is a fact that because K_1 is an sNE and ESS, it not only cannot be invaded by any of the other pure strategies K_2 to K_4, but none of the mixed strategies can invade it either. So the uniqueness of tracking as an ESS (sNE) is completely general over theories of knowledge that see knowledge as true belief plus a further condition. There may be conditions that are not the tracking conditions but do imply them, and those conditions would also count as an ESS (sNE). But since it would be only in virtue of implying the tracking conditions that they guaranteed that stability, it is still the tracking that confers the value that an ESS and an sNE bring.[9]

When speaking of strategies in these games I have used locutions in which the players play a strategy or follow a rule, because they are less misleading than talk about choosing options, since the former can be done unconsciously. When a subject needs to know p she rarely just finds herself choosing between her tracking option and other options that could lead to error. There are matters on which a human being does come naturally equipped with a tracking ability—for example, her eyes can track whether there is a tiger in front of her—but in those cases we would not think of her as, and she would not typically be, choosing to use her eyes. Her doing that is automatic. In nonperceptual cases the subject often would choose the tracking option if she could, but it is not just there for the choosing. A scientist would have to build a hadron collider in order to set up a set of messages that distinguish, for example, the existence and nonexistence of the particle of interest. Journalists, and many other types of knowledge seekers, have to do work to set up a tracking relation with the truth of interest. Even the locutions of "playing a strategy" and "following a rule" can be misleading for the nonauto-matic cases to the extent that they suggest a mere decision.

We can think of these strategies in the True Belief Game as rules that with repeated achievement and use can become dispositions of subjects, but the notion of following a rule has an ambiguity worth clarifying here too. One may follow a rule that corresponds to one of the strategies by using its response types in a given round of play. However, that does not

---

[8] If there are no such probabilities for how fulfillment of a proposed requirement for knowledge would make you do in the task of believing p when p and not believing when $-p$, then the requirement has no truth connection at all, and so could not invade our tracking-based strategy anyway.

[9] Tracking with closure (Roush 2005) is weaker than tracking because it allows one also to know p in virtue merely of tracking some q that one knows implies p. On that view, knowing p would not imply one had an ESS for p. However, knowing p would imply that either one had an ESS for p or one had an ESS for some q that implies p.

correspond to tracking; tracking corresponds to playing or following that strategy or rule as a habit or disposition with respect to p—it requires that a player is of a type.

How a subject can or does get himself into the position of having strategies available is not a concern of this game-theoretic analysis. One may stumble into doing a tracking type play, one may be automatically disposed to it, one may make a herculean effort to get to be able to choose and commit to use it. Also, it is the same if you had a strategy available and didn't use it or didn't use it because you didn't have it. What matters to the outcomes and stability properties is only whether one acts in accord with a particular rule, and whether one is or becomes disposed to do so.

This restriction of attention does not undermine the relevance of these results to epistemology, however, since how one gets oneself into the state of knowledge, despite being a question of central interest in epistemology, is not per se a topic relevant to the questions of whether one is in that state or not and what is required for counting as being in that state. A particular view of the criteria for being in that state may stipulate that how one got there matters to whether one is in the state—genetic views of what knowledge is, like process reliabilism, do that—but that is a choice of a particular theory, not a requirement for having an answer at all to the question of what knowledge is. Counterfactual theories care only about what your properties now say that you would believe in an alternate situation or whether what you would believe in an alternate situation would be true. What many internalist justification views of knowledge care about is whether you currently have reasons available. The concern in the value problem too, just like the concern of the True Belief Game, is not how you got to your knowledge, or your capacity to have strategies at all, but what it is you now have by being there.

What does it follow that you have on the tracking view when you have achieved knowledge? What follows is whatever follows from having true belief in p and the strong disposition to follow the strategy that is the unique strict Nash Equilibrium and the unique ESS in the True Belief Game for p. The dominance of your strategy—its being an sNE—brings a number of properties with it. It brings generality over rounds of play: you will win in every round of the True Belief Game (except those few stages where you play out of character—your disposition to follow K_1 is not assumed perfect). You will always get a higher payoff than any other strategy could get you. Winning a round doesn't necessarily give you knowledge, because it does not necessarily bring you belief. The state of the world may be p or $-p$, and if it is $-p$ then your winning will come from the clause of your strategy that makes you *not* believe in such circumstances. However, this absence of belief is valuable too, we assumed, the most valuable thing you could do given that state of the world.

Having knowledge, on the tracking view, implies having a true belief and along with it a disposition that would make you have the epistemic state—

belief or nonbelief—that is most valuable given the state of the world, in almost every round, were you to play the True Belief Game an infinite number of times. This security may sound so simple as to be trivial, but its power lies in the fact that in an infinite number of rounds of play the game could have an infinite number of different manifestations. As long as the payoff structure holds constant, having your sNE means you will win in all the remotely probable manifestations. What kinds of variation could there be? It could be that in the actual case of your true belief that a particular road is the road to Larissa you are not having a discussion with a sophist, but in another round though it still is the road to Larissa you are also stopped along the way by a wily, argumentative guy. If in addition to having a true belief you are a K_1 type of subject on the matter of this road, then you would believe it both were you to be talking to a sophist and were you not. If the sophist were to give you a bad argument that it is not the road to Larissa then you, the K_1 type, wouldn't give up your belief. Being K_1 implies that somehow or other you would know better. The subject with a mere true belief would be a sitting duck for the sophistical trick.

In this case the truth value of p did not vary from the actual, but the circumstances did. There could also be a variation in truth value of p over different rounds. Though this is in fact now the road to Larissa, there could be a round of the game where it isn't. If a person merely has a true belief that this is the road, then nothing follows about whether she would pick up on the circumstance where it wasn't. There may be signs and trustworthy authorities to tell people where the road goes instead, but being a mere true believer gives no assurance that you would pick up on them. The K_1 type of subject, by contrast, is prepared to have an appropriate belief state even if there turns out to be road work.

The counterfactual properties that flow from living in a strict Nash Equilibrium give the subject who has knowledge preparedness for all probable circumstances and changes in the truth value of p. However, we have said that properties of the history of a belief cannot save a view of knowledge from the swamping problem, so one might wonder how properties of the subject's potential future could; why isn't the problem symmetric in time? The basic reason is that time flows in one direction. Everything from the past that is relevant to epistemic success now has had its chance to be taken into account in the actual present belief; what the future may hold cannot have been captured already in that belief. And this is not only because the future has not actually happened yet but also because what does happen in it will not be determined exclusively by the subject's currently believing or even by that belief's truth. There are a million other present and future conditions not determined by this belief or its truth. What the subject will do in response to those circumstances is also not determined by her actually having a true belief now. Her having a disposition to a strategy in the True Belief Game does (probabilistically) determine this.

Being K_1 now does add something now that is identifiable and not redundant with merely having a true belief, but we can still ask whether that thing is valuable. This comes down to the question of whether *preparedness* is valuable, since what being type K_1 now gives is robustness of epistemic success against future contingencies. It seems to me undeniable that preparedness has added value, since denying it would require denying that true belief has extrinsic value at any time before it is actually being used. If the value of a true belief is that it aids you in achieving something else, then it is valuable at the times when it is actually aiding you, but it would not be valuable at any earlier time unless we supposed that the *potential* for aiding was also valuable. If we denied that preparedness is valuable, then the true belief about which is the road to Larissa wouldn't be valuable at all except at those times when we were actually walking on the road with the intention of going to Larissa. We don't think that, so preparedness is valuable.[10] The preparedness that K_1, or tracking, brings is not redundant with the potential a mere true belief brings, since whatever success the current mere true belief might give the subject in the future will be dominated in payoffs by what the tracking true belief brings. Thus, tracking is both additional and valuable.

The security that the dynamical stability property of tracking brings is a form of persistence over time, and the value of knowledge over true belief has been associated by some, including Socrates, with persistence. The difference between the current view and the others is in what is expected to persist when you have knowledge of p. Socrates compared true belief with the statues of Daedalus: magnificent creations, but they run away if not tied down. Mere true belief will not stay around long either, said Socrates, unless it is tethered, in his view by working out the reason (Plato, *Meno*, 97d–98a). Timothy Williamson has fleshed this idea out by arguing that knowledge is literally more persistent than true belief in part because it is less susceptible to rational undermining (Williamson 2000, 79). Kvanvig (2003, 13–20) argues against this, and despairs of the prospects for any persistence view of the added value of knowledge. What all of this misses is that it is not true belief, or knowledge, that will persist in virtue of one's having knowledge. It is not even appropriate for a theory of knowledge to imply that knowledge of any contingent truth persists, or is likely to persist, because the *truth* of a contingent truth cannot be expected to persist.[11] Roads change, tigers come back, and I'm afraid that chocolate shops sometimes go out of business.

---

[10] There are other epistemological cases of the added value of preparedness. Many internalists think that having an argument consciously accessible is valuable, but even if an argument that is being used is valuable an argument that is merely accessible and is not actually being used wouldn't have value if preparedness had no value.

[11] The truth of necessary truths does, of course, persist, so tracking is not the appropriate kind of responsiveness to have to necessary truths. The appropriate kind is proposed and defended in Roush 2005 (134–47).

If you have a list of truths about where the chocolate shops are in town, an example due to Kvanvig, then you have something valuable (if you like chocolate), but you do not, he points out, have something more valuable if you have the intersection of this list with a list of where the chocolate shops are *likely* to be. However, I do have something more valuable if I have in addition to a list of truths about where the chocolate shops are a responsiveness to their probabilities of going out of business over the next month, quarter, and year; I will be disposed now to try them differentially in the future in a way that ensures more success and efficiency in getting my chocolate. If a mere true belief that I have now, before deciding in which direction to walk to get my chocolate, is valuable to me because it raises my chances of getting chocolate, then my having now a responsiveness of my beliefs to future closings is of value too, and it is evidently not redundant.

What persists over time for a K_1 type of believer in p is not knowledge or belief that p, there is a chocolate shop at a certain place, but *appropriateness* of epistemic state—belief or nonbelief in p—over time and changing circumstances. The appropriateness is cashed out in my getting the highest payoff a player could get no matter the state of the world, p or $-p$. And payoffs, of course, are not restricted to nonessential pleasures. They may be food versus no food, shelter or health care versus none; they may be any of the goods, services, and cooperative relationships that are relevant to survival. Provided the payoff structure of the game remains the same, the knower type gets the highest payoff that any type could get, and is highly likely to do so in the future, in every round of play; baldly put, the K_1 type of believer is more likely to survive and flourish. Since persistence and advantage are intuitive features of the value of knowledge, remarked upon by philosophers of various persuasions, the fact that K_1 ensures a sensible version of them is a point in favor of tracking as a theory of what knowledge is.

The value of knowledge I have been discussing so far flows from K_1 being a strict Nash Equilibrium and what follows from this about the fate of an individual who is a tracker in comparison to individuals of other types. That K_1 is an ESS brings in a further dimension that seems to have explanatory force when applied to human populations over time, even historical and evolutionary time. An ESS type has a resistance to extinction. In our game this means that, if payoffs remain the same over generations of play, then a widespread subpopulation of knowers of p cannot be eliminated—as a type—by any small population of any style of ignoramus type. This holds, recall, no matter what the dynamics of interaction or variations in circumstance.

One might think of this in connection with the ideas that education brings greater success, that the truth will prevail, that an educated population will be less likely to enact policies that will lead to its own destruction, that though the meek may inherit the earth the ignorant will not. We might associate it with hope that Karl Rove was wrong to express disdain for the "reality-based community," people who "believe that solutions emerge from your judicious study of discernible reality," that he was wrong to think "That's not the way the world really works anymore. We're an empire now, and when we act, we create our own reality" (qtd. Suskind 2004). We might think of the hope that flat-out false beliefs will not determine the outcomes of elections, or of Congressional policy votes.

These hopes are often thought of as naïve, but the ESS property of knowledge gives them some basis. A subpopulation of organisms of a type which has tiger detection in an environment that has tigers that like to eat them, and which is a large fraction of the total population of such organisms, will never be outcompeted as a type—that is, eliminated as a proportion of the entire population—by a small subpopulation of organisms that lack tiger detection. This case is clean because the connection between accurate representation of reality and positive relative payoff is direct, and the payoffs are not imagined as changing in the course of repeated play. In such cases knowledge, or its counterparts involving tracking via representations more primitive than belief, will have a ratchet effect on the evolution of species. If a knowledge-bearing type of organism arises and goes to fixation—"everybody" knows—then no ignorant variant within the species can stop the future survival of the knower type. It is worth emphasizing that an ESS ensures nothing about the fate of individuals—that is a concern we dealt with earlier in the context of an sNE. In interpreting what it means to say that the knower type will not be driven to extinction, it seems most natural to say that *knowledge* is what will survive; this is not quite true, but close: what will survive is belief states appropriate to the potentially changing states of the world with respect to p, which of course requires lack of belief in p when p is false. This will survive despite potentially being borne by different individuals in each round of play.

In human history and culture, the knower type does not always prevail, not even a large population of knower types. We can explain the consistency of this with a tracking view of knowledge by the fact that human beings have capacities that can lead to failure of the conditions for the ESS outcome. One such condition is that the payoff structure of the game remain constant over repeated play; repeated play, in political contexts for example, often changes the payoffs. If one despises Karl Rove, then it will be at least partly because of his evaluative claim that we are permitted to change reality however we please if we have enough power. But he was right in the factual claim that human actions can change reality, not only in the obvious sense that occurs when we build bridges, but also because human actions can change payoff structures.

For example, in a pitched public battle over a government policy, people with enough resources and cleverness can blanket the media with messages psychologically well crafted to convince the populace of outright falsehoods, with the goal of making them proponents or opponents of the policy. The best payoff for a member of Congress deliberating over whether to vote for the policy might have been determined by whether it truly would improve the lives of his constituents, because after all even if his only concern was reelection his constituents' well-being would be the thing that determined whether they voted for him—right? However, if the election is soon and the consequences of the policy will emerge more slowly, and the member of Congress doesn't have the resources to counter sufficiently the falsehoods that his constituents have come to believe through aggressive advertising, then what is truly in their best interest will not be determining their vote in the next election. The member of Congress now has the highest payoff from voting in the direction of policy that will not help his constituents.

For any p a true belief in which is valuable, and so on, the knower type, if in sufficient numbers, will survive as long as conditions relevant to payoffs remain the same. We can see through empirical examples that one condition for their remaining the same is that there be no relevant deception. However, this amounts to saying that the mere *having* of knowledge by many people will not prevent the damaging consequences of deception, and it is not surprising that overcoming the effects of outright obfuscation on a system of interaction will require not just knowledge but also countermeasures.

It is an understatement to say that among human beings the applicability of the ESS result will not be universal; it is a result in an idealization that has the kind of usefulness such tools bring. We have seen that the idealization implies that even perfect knowledge does not alone give us everything we need epistemically, but we already knew that. We also know that knowledge gives a lot, and the tracking view of what knowledge is provides a simple and powerful picture for explaining what and how that is.

Appendix

From

$$Pr(M_1/p) = \text{very high}$$

$$Pr(B(p)M_1) = \text{very high} \qquad (*)$$

and

$$Pr(B(p)/p.M_1) = Pr(B(p)/M_1)$$

We wish to derive

$$Pr(B(p)/p) \text{ is high} \qquad (\dagger)$$

By $*$,

$$Pr(M_1.p) \approx Pr(p)$$

$$Pr(B(p).M_1) \approx Pr(M_1) \qquad (!)$$

$$Pr(B(p)/p.M_1) = Pr(B(p)/M_1)$$

implies

$$Pr(B(p).p.M_1)/Pr(p.M_1) = Pr(B(p).M_1)/Pr(M_1)$$

By !,

$$Pr(B(p).p.M_1)/Pr(p.M_1) \approx 1$$

which implies

$$Pr(B(p)/p.M_1) \approx 1$$

But by !

$$Pr(M_1.p) \approx Pr(p)$$

so

$$Pr(B(p)/p) \approx 1.$$

Two approximate equalities were used in the derivation of the final inequality, so there are two sources damping down the final correlation, and so it may be lower. This is why I claim only that high tracking correlations come from very high Signaling System correlations rather than that very high come from very high. The derivation for the second half of strategy K_I is analogous. The nature of probabilistic tracking is discussed in Roush 2005; lower bounds on the loss of correlation over tracking links are shown in chapter 5, where the links are chains of evidence. Both parts of a strategy like K_I are actually tracking conditions, though not the specific ones about p and $B(p)$ that we use to analyze knowledge.

Acknowledgments

This work was supported by NSF Award No. SES-0823418. Special thanks to Jason Alexander for helpful discussions, and to Brian Skyrms for teaching me most of what I know about game theory.

References

Floridi, Luciano. 2005. "Consciousness, Agents, and the Knowledge Game." *Minds and Machines* 15:415–44.
Goldman, Alvin. 1979. "What Is Justified Belief?" In *Justification and Knowledge*, edited by George Pappas, 1–23. Dordrecht: D. Reidel.

Kvanvig, Jonathan. 2003. *The Value of Knowledge and the Pursuit of Understanding*. Cambridge: Cambridge University Press.

Lewis, David K. 1969. *Convention*. Cambridge, Mass.: Harvard University Press.

Roush, Sherrilyn. 2005. *Tracking Truth: Knowledge, Evidence, and Science*. Oxford: Oxford University Press.

Suskind, Ron. 2004. "Faith, Certainty, and the Presidency of George W. Bush." *New York Times Magazine*, October 17.

Swinburne, Richard. 1999. *Providence and the Problem of Evil*. Oxford: Oxford University Press.

———. 2000. *Epistemic Justification*. Oxford: Oxford University Press.

Williamson, Timothy. 2000. *Knowledge and Its Limits*. Oxford: Oxford University Press.

3

KNOWLEDGE TRANSMISSIBILITY AND PLURALISTIC IGNORANCE: A FIRST STAB

VINCENT F. HENDRICKS

In his contribution to *Philosophy of Computing and Information: Five Questions*, Luciano Floridi explains why he was initially drawn to informational issues:

> The second reason was related to what I like to describe as *methodological minimalism*. I was looking for a more "impoverished" approach, which could allow me to work on more elementary concepts, less worn down by centuries of speculation, and more easily manageable. It seemed that, if one could have any hope of answering difficult questions about complicated issues concerning knowledge, meaning, the mind, the nature of reality or morality, it made sense to try to tackle them at the lowest and less committed level at which one could possibly work. Informational and computational ideas provided such a minimalist approach. To give a concrete example, my interest in artificial agents was motivated by the classic idea that less is more. This is still not very popular among philosophers, who seem to be too much in love with the human subject, his psychology and idiosyncrasies.
>
> (Floridi 2008, 94)

Floridi's artificial agents may interact—sometimes conducive to inquiry and information aggregation, other times obstructing the very same. Especially social psychologists and economists have for a while, both theoretically, using computational tools from observational learning theory, and empirically, through experiments, scrutinized the mechanics of information obstruction while agents interact. Less is more indeed, and techniques from formal epistemology may be wired into these studies, providing both models and sometimes resolution. Here is a first stab using an "elementary concept" from epistemic logic, knowledge transmissibility, to tackle pluralistic ignorance and the informational issues related to this phenomenon.

1. Pluralistic Ignorance

Pluralistic ignorance may appear when a group of decision makers have to act or believe at the same time when given a public signal (Bikhchandani, Hirshleifer, and Welch 1998). When asking a group of new students what they thought was difficult in the reading material handed out for today's lecture, chances are good that nobody will signal their comprehension problems. While deciding whether to flag ignorance or not, the individual student first observes whether the other students encountered the same problem of text comprehension. When all the students are doing this at the same time, the public signal becomes that nobody found the text difficult. Thus, in order not to hurt his standing, nobody signals ignorance. Everybody refrains from acting on personal information exactly because nobody acts immediately on her personal information. In sum, the danger for pluralistic ignorance arises when the individual decision maker in a group lacks the necessary information for solving a problem at hand, and thus observes others hoping for more information. When everybody else does the same, everybody observes the lack of reaction and is consequently led to erroneous beliefs. This emperor's new clothes mechanism is seen in bystander effects where people are more likely to intervene in an emergency when alone rather than in the presence of others, and it is widely used in sales campaigns and policy making. However, pluralistic ignorance is also fragile, as the phenomenon only stands firmly as long as nobody cries out. And so we have the question, in sync with Floridi's methodological minimalism: What is the epistemic nature of information leading to bursting the bubble of ignorance?

2. Modal Operator Epistemology

The basic formal setup is modal operator epistemology (Hendricks 2001, 2003, 2007), which is the cocktail obtained by mixing formal learning theory (Kelly 1996) and epistemic logic in order to study the formal properties of limiting convergence knowledge:

- An evidence stream ε is an ω-sequence of natural numbers, i.e., $\varepsilon \in \omega^{\omega}$.
- A possible world has the form (ε,n) such that $\varepsilon \in \omega^{\omega}$ and the state-coordinate $n \in \omega$.
- The set of all possible worlds $W = \{(\varepsilon,n)|\varepsilon \in \omega^{\omega}, n \in \omega\}$.
- $\varepsilon|n$ denotes the finite initial segment of evidence stream ε of length n.
- Define $\omega^{<\omega}$ to be the set of all finite initial segments of elements in ω.
- Let $(\varepsilon|n)$ denote the set of all infinite evidence streams that extends $\varepsilon|n$.
- The set of possible worlds in the fan, i.e., background knowledge, is defined as

$$[\varepsilon|n] = (\varepsilon|n) \times \omega.$$

A possible world in the current setup is a significantly different entity from the classical conception. On the traditional view, a possible world is

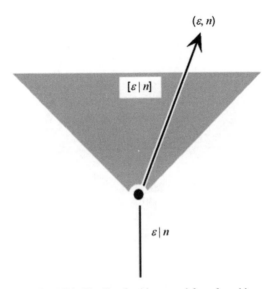

FIGURE 1. Handle of evidence and fan of worlds

not a composite but rather an entity complete in its spatiotemporal history. Here a possible world is a pair consisting of an evidence stream or tape and a state-coordinate allowing one to point to specific entries in the tape, specify all the evidence seen up to a certain point in the inquiry process, quantify over all or some evidence in the evidence tape, and so on. Thus, what is observed "right now" is simply the content of a cell in the evidence tape at the specific time "now" that the state-coordinate determines.

2.1. Hypotheses

Hypotheses will be identified with sets of possible worlds. Define the set of all simple empirical hypotheses as

$$\mathscr{H} = P(\omega^\omega \times \omega).$$

A hypothesis h is said to be *true* in world (ε, n) iff

$$(\varepsilon, n) \in h \text{ and } \forall l \in \omega : (\varepsilon, n + l) \in h.$$

Truth requires identification and inclusion of the actual world (ε, n) in the hypothesis for all possible future states of inquiry.

3. Agents and Inquiry Methods

An inquiry method (or agent) may be one either of discovery or of assessment. A discovery method δ is a function from finite initial

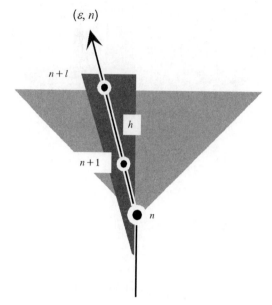

FIGURE 2. Truth of a hypothesis h in a possible world (ε, n)

segments of evidence to hypotheses, that is:

$$\delta : \omega^{<\omega} \rightarrow \mathcal{H}. \tag{1}$$

The convergence modulus for a *discovery* method (abbreviated *cm*) where μ is the least search operator also known as minimalization is given by:

$$cm(\delta, h, (\varepsilon, n)) = \mu k \forall n' \geq k \forall (\tau, n') \in [\varepsilon|n] : \delta(\tau|n') \subseteq h.$$

An assessment method α is a function from finite initial segments of evidence and hypotheses to true/false, that is:

$$\alpha : \omega^{<\omega} \times \mathcal{H} \rightarrow \{0, 1\} \tag{2}$$

The convergence modulus for an assessment is defined in the following way:

$$cm(\alpha, h, (\varepsilon, n)) = \mu k \geq n, \forall n' \geq k, \forall (\tau, n') \in [\varepsilon|n] : \alpha(h, \varepsilon|n) = \alpha(h, \tau|n').$$

3.1. Knowledge Based on Discovery

Limiting knowledge for a discovering agent in a possible world requires that (1) the hypothesis is true, (2a) the agent in the limit settles for a conjecture entailing the hypothesis, and (2b) the agent does so infallibly in the sense of entailment of the hypothesis by the observed evidence.

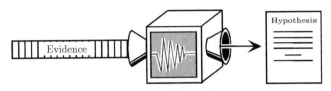

FIGURE 3. A discovery method δ

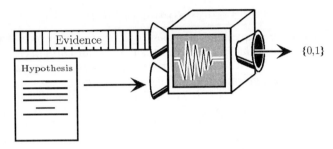

FIGURE 4. An assessment method α

Formally, this condition for limiting convergent discovered knowledge amounts to:

(ε,n) validates $K_\delta h$ iff:

1. $(\varepsilon,n) \in h$ and $\forall l \in \omega$: $(\varepsilon,n+l) \in h$,
2. $\forall n' \geq n$, $\forall(\tau, n') \in [\varepsilon|n]$:

    (a) $\delta(\tau|n') \subseteq h$.
    (b) $(\tau|n) \in \delta(\tau|n')$

The discovery method may additionally be subject to certain agendas (methodological recommendations) like perfect memory, consistency, infallibility, and so forth (Hendricks 2001), but they are of little concern here.

3.2. Knowledge Based on Assessment

Limiting knowledge for an assessing agent in a possible world requires that (1) the hypothesis is true and (2) that the agent in the limit decides the hypothesis in question. This formally comes to:

(ε,n) validates $K_\alpha h$ iff:

1. $(\varepsilon,n) \in h$ and $\forall l \in \omega$:$(\varepsilon,n+l) \in h$,
2. α decides h in the limit in $[\varepsilon|n]$:

    (a) if $(\varepsilon,n) \in h$ and $\forall l \in \omega$:$(\varepsilon,n+l) \in h$ then
       $\exists k \geq n, \forall n' \geq k$, $\forall(\tau, n') \in [\varepsilon|n]$:$\alpha(h,\tau|n') = 1$,

(b) if $(\varepsilon,n) \notin h$ or $\exists l \in \omega : (\varepsilon,n+l) \notin h$ then
$\exists k \geq n, \forall n' \geq k,\ \forall (\tau,\ n') \in [\varepsilon|n] : \alpha(h,\tau|n') = 0.$

4. Multimodal Systems

The above set-theoretical characterization of inquiry lends itself to a multimodal logic. The modal language $\mathscr{L}$ is defined accordingly:

$$A ::= a | A \wedge B | \neg A | K_\delta A | K_\alpha A | [A!]B | I_\delta A | I_\alpha A$$

where $[A!]B$ is the public announcement operator to be read as "after it has been announced that A, then B is the case" (van Ditmarsch, van der Hoek, and Kooi 2007; Roy 2008) and $I_\delta A$ is the ignorance operator to be read as "agent δ is ignorant of A" (van der Hoek and Lomuscio 2004). In a relational Kripke-model, $[A!]B$ means that after deleting all situations in which A does not hold in the original model, B holds. Put differently, $[A!]B$ takes one from the initial model to a new model; that is, the accessibility relation is between worlds in different models. $I_\delta A$ says that δ is ignorant of A if there exist two distinct situations or possible worlds—one in which A is true, and another one in which A is false. Operators for alethic as well as tense may also be added to L (see Hendricks 2001 and 2007 for details).

Consider this model:

$\mathbb{M} = <\mathscr{W},\ \varphi,\ \delta,\ \alpha>$ consists of:

1. A nonempty set of possible worlds $\mathscr{W}$,
2. A denotation function φ: *Proposition Letters* $\rightarrow P(\mathscr{W})$, i.e., $\varphi(a) \subseteq \mathscr{W}$.
3. Inquiry methods:

   (a) $\delta : \omega^{<\omega} \rightarrow P(\mathscr{W})$
   (b) $\alpha : \omega^{<\omega} \times \mathscr{H} \rightarrow \{0,1\}$

If a is a propositional variable, then semantical correctness may be defined such that $(\varepsilon,n) \Vdash_M a$ iff $(\varepsilon,n) \in \varphi_M(a)$. Next, the semantical meaning of an arbitrary formula A in the language may be defined as $[A]_M = \{(\varepsilon,n) | (\varepsilon,n) \Vdash A\}$. The truth conditions for the Boolean connectives and remaining operators may be defined as follows:

Let $\varphi_{M,(\varepsilon,n)}(A)$ denote the truth-value in (ε,n) of a modal formula A given M defined by recursion through the following clauses:

1. $\varphi_{M,(\varepsilon,n)}(a) = 1$ iff $(\varepsilon,n) \in \varphi(a)$ and $\forall l \in \omega : (\varepsilon,n+l) \in \varphi(a)$ for all propositional variables a, b, c, ...
2. $\varphi_{M,(\varepsilon,n)}(\neg A) = 1$ iff $\varphi_{M,(\varepsilon,n)}(A) = 0.$
3. $\varphi_{M,(\varepsilon,n)}(A \wedge B) = 1$ iff both $\varphi_{M,(\varepsilon,n)}(A) = 1$ and $\varphi_{M,(\varepsilon,n)}(B) = 1$; otherwise $\varphi_{M,(\varepsilon,n)}(A \wedge B) = 0.$

4. $\varphi_{M,(\varepsilon,n)}(K_\delta A) = 1$ iff

   (a) $(\varepsilon,n) \in [A]_M$ and $\forall l \in \omega:(\varepsilon,n+l) \in [A]_M$,
   (b) $\forall n' \geq n,\ \forall (\tau,\ n') \in [\varepsilon|n]:\delta(\tau|n') \subseteq [A]_M$.

5. $\varphi_{M,(\varepsilon,n)}([A!])B) = 1$ iff $\varphi_{M,(\varepsilon,n)}(A) = 1$ implies $\varphi_{M,(\varepsilon,n)|A}(B) = 1$.
6. $\varphi_{M,(\varepsilon,n)}(I_\Xi A) = 1$ iff $\exists(\tau,m)\exists(\mu,m') \in [\varepsilon|n]:\tau|n = \mu|n$, and
   $\varphi_{M,(\tau,m)}(A) = 1$ and $\varphi_{M,(\mu,m')}(\neg A) = 1$ for $\Xi \in \{\delta,\alpha\}$.

Model $\mathbb{M}$-subscript will be suppressed when it is clear from context. Knowledge based on assessment is omitted from the definition above for reasons of brevity. Note that item 5 introduces a model-restriction $(\varepsilon,n)|A$ similar to the one described in Roy 2008. Item 6 follows the definition of ignorance defined in van der Hoek and Lomuscio 2004. Accessibility is no longer defined as a binary relation on points. Rather, two worlds (τ,m) and (μ,m') are accessible from each other if they have the same handle, i.e. $\tau|n = \mu|n$ relative to $[\varepsilon,n]$.

5. Knowledge Transmissibility

Already in *Knowledge and Belief* (1962) Hintikka considered whether

$$K_\beta K_\gamma A \rightarrow K_\beta A \qquad (3)$$

is valid (or self-sustainable, in Hintikka's terminology) for arbitrary agents β,γ. Now (3) is simply an iterated version of Axiom **T** for different agents, and so long as β,γ index the same accessibility relation the claim is straightforward to demonstrate. From an active agent perspective the claim is less obvious.

The reason is agenda-driven or methodological. Inquiry methods β,γ may—or may not—be of the same type. If knowledge is subsequently defined either on discovery or on assessment, then (3) is not immediately valid unless discovery and assessment methods can "mimic" or induce each others' behavior in the following way:

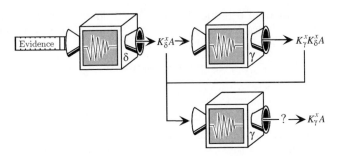

FIGURE 5. Knowledge transmissibility and inquiry methods

THEOREM 1. If a discovery method δ discovers h in a possible world (ε,n) in the limit, then there exists a limiting assessment method α which verifies h in (ε,n) in the limit.

Proof. Assume that δ discovers h in (ε,n) in the limit and let

$$cm(\delta, h, (\varepsilon, n))$$

be its convergence modulus. Define α in the following way:

$$\alpha(h, \varepsilon|n) = 1 \text{ iff } \delta(\varepsilon|n) \subseteq h.$$

It is clear that if $n' \geq cm(\delta,h,[\varepsilon|n])$, then for all $(\tau,n') \in [\varepsilon|n]:\delta(\tau,n') \subseteq h$. Consequently $\alpha(h, \tau,n') = 1$ and therefore

$$cm(\alpha, h, (\varepsilon, n)) \leq cm(\delta, h, (\varepsilon, n)).$$

Similarly, but conversely: □

THEOREM 2. If an assessment method α verifies h in (ε,n) in the limit, then there exists a limiting discovery method δ which discovers h in (ε,n) in the limit.

Proof. Similar construction as in proof of theorem 1.   □

Using inducement it is easily shown that:

THEOREM 3. $K_\delta A \leftrightarrow K_\alpha A$.

Proof. ($\rightarrow$) Show that $(\varepsilon,n) \vdash K_\delta A \rightarrow K_\alpha A$. Then

1. $(\varepsilon,n) \in [A]$ and $\forall l \in \omega : (\varepsilon,n+l) \in [A]$.
2. $\forall n' \geq n,\ \forall(\tau,\ n') \in [\varepsilon|n]$

   (a) $\delta(\tau|n') \subseteq [A]$,
   (b) $(\tau,n) \in \delta(\tau|n')$.

Let discovery method δ induce its assessment correlate α in accordance with theorem 1. Since $(\varepsilon,n) \in [A]$ and $\forall l \in \omega : (\varepsilon,n+l) \in [A]$ by (1) construct the assessment correlate in such a way that

$$\forall n' \geq n, \forall (\tau, n') \in [\varepsilon|n] : \alpha([A], \tau|n') = 1 \leftrightarrow \delta(\tau|n') \subseteq [A].$$

But by (2) $\forall n' \geq n, \forall(\tau, n') \in [\varepsilon|n] : \delta(\tau|n') \subseteq [A]$ so

$$\exists k \geq n, \forall n' \geq k, \forall(\tau, n') \in [\varepsilon|n] : \alpha([A], \tau|n') = 1$$

which amounts to the definition of knowledge based on assessment. ($\leftarrow$) Show that $(\varepsilon,n) \vdash K_\alpha A \rightarrow K_\delta A$. This proof is similar to ($\rightarrow$) observing theorem 2.   □

In Hendricks 2001 it is proved that knowledge based on discovery and assessment under suitable methodological recommendations validates **S4**. Hintikka's original knowledge transmissibility thesis (3) may be easily proved now, since theorems 1, 2, and 3 provide the assurance that a discovery method can do whatever an assessment method can do, and vice versa:

$$K_\delta K_\alpha A \rightarrow K_\delta A$$

(i)	$K_\alpha A \rightarrow A$	Axiom **T**
(ii)	$K_\delta(K_\alpha A \rightarrow A)$	(i), (Nec. K_δ)
(iii)	$K_\delta(K_\alpha A \rightarrow A) \rightarrow (K_\delta K_\alpha A \rightarrow K_\delta A)$	(ii), Axiom **K**
(iv)	$K_\delta K_\alpha A \rightarrow K_\delta A$	(ii), (iii), (MP)

In the literature one often encounters defences for epistemic norms derived from (3) like:

CONJECTURE 1. If δ knows that A, then δ has a certain degree of evidence for A, or δ has had A transmitted from γ who has evidence for A.

This may indeed be so, but "evidence for A" is essentially an issue related to how the evidence or information has been collected—thus an agenda concern, or a methodological concern, of how inquiry methods for knowledge acquisition interact.

6. Knowledge Transmissibility and Public Announcement

In order to study knowledge transmissibility and pluralistic ignorance, the relation between transmissibility and public announcement has first to be uncovered. Here is a set of theorems relating knowledge transmissibility to public announcement. Proofs are omitted but may be found in Hendricks forthcoming.

Let there be given a finite set of discovery agents $\Delta = \{\delta_1, \delta_2, \delta_3, \ldots, \delta_n\}$ and a finite set of assessment agents $\Lambda = \{\alpha_1, \alpha_2, \alpha_3, \ldots, \alpha_n\}$, and assume that all agents in Δ, Λ have **S4**-knowledge. Now it may be shown that (3) holds for agents of different types:

THEOREM 4. $\forall \delta_i \in \Delta : K_{\delta_i} K_\alpha A \rightarrow K_{\delta_i} A$ if theorem 1 holds for $\Delta = \{\delta_1, \delta_2, \delta_3, \ldots, \delta_n\}$.

THEOREM 5. $\forall \alpha_i \in \Lambda : K_{\alpha_i} K_\delta A \rightarrow K_{\alpha_i} A$ if theorem 2 holds for $\Lambda = \{\alpha_1, \alpha_2, \alpha_3, \ldots, \alpha_n\}$.

The next two theorems show that the axiom relating public announcement to knowledge given the standard axiomatization of public announcement logic with common knowledge (van Ditmarsch, van der Hoek, and Kooi 2007) holds for knowledge based on discovery and knowledge based on assessment. This is a variation of the original knowledge prediction axiom, which states that "some a knows B after an announcement A iff (if A is true, a knows that after the announcement of A, B will be the case)":

THEOREM 6. $\forall \delta_i \in \Delta : [K_\alpha A!]K_{\delta_i}B \leftrightarrow (K_\alpha A \rightarrow (K_{\delta_i}K_\alpha A \rightarrow K_{\delta_i}[K_\alpha A!]B))$ if theorem 2 holds for $\Delta = \{\delta_1, \delta_2, \delta_3, \ldots, \delta_n\}$.

THEOREM 7. $\forall \alpha_i \in \Lambda : [K_\delta A!]K_{\alpha_i}B \leftrightarrow (K_\delta A \rightarrow (K_{\alpha_i}K_\delta A \rightarrow K_{\alpha_i}[K_\delta A!]B))$ if theorem 3 holds for $\Lambda = \{\alpha_1, \alpha_2, \alpha_3, \ldots, \alpha_n\}$.

7. Knowledge Transmissibility and Pluralistic Ignorance

Recall that pluralistic ignorance may appear when the individuals in a group lack the required information for solving a current problem, and thus observe others hoping for more information. When everybody else does the same, everybody observes the lack of reaction and is consequently led to erroneous beliefs. A simple situation of pluralistic ignorance arises in Hans Christian Andersen's fable *The Emperor's New Clothes*, in which a small boy dispels the spell of pluralistic ignorance by publicly announcing the lack of garments on the emperor. To model this particular situation, assume:

1. A finite set of ignorant agents either based on discovery of assessment or both:

   (a) $\Delta = \{\delta_1, \delta_2, \delta_3, \ldots, \delta_n\}$
   (b) $\Lambda = \{\alpha_1, \alpha_2, \alpha_3, \ldots, \alpha_n\}$

2. A public announcement such that after it has been announced that A, then B is the case:

   (a) $[A!]B$.

Assume furthermore the existence of

1. At least one knowing agent based on either discovery or assessment:

   (a) $K_\alpha A$,
   (b) $K_\delta A$,

2. Inducement theorems 1 and 2.

Pluralistic ignorance based on discovery or assessment may established in the following way:

3. Discovery: $\forall \delta_i \in \Delta : [C!]I_{\delta_i}A$,
4. Assessment: $\forall \alpha_i \in \Lambda : [C!]I_{\alpha_i}A$,

where C is the public signal (announcement) leading all agents in Δ or Λ to ignorance of A. The airport announces C: "Some flights are delayed," and every $\delta_i \in \Delta$ becomes ignorant as to A: "My flight is delayed" or $\neg A$: "My flight is not delayed." Or the teacher publicly announces C: "This exercise is tricky," and every student $\forall \alpha_i \in \Lambda$ becomes ignorant as to A: "I got it right" or $\neg A$: "I got it wrong."

Suppose there is a knowing agent (the little boy in *The Emperor's New Clothes*) such that

5. $K_\alpha A$ for bullet 3,
6. $K_\delta A$ for bullet 4.

Now it is possible to prove the following two theorems.

THEOREM 8. $\forall \delta_i \in \Delta : I_{\delta_i} A \wedge ([K_\alpha A!](A \wedge K_{\delta_i} K_\alpha A)) \rightarrow [K_\alpha A!]K_{\delta_i} A$ if theorem 1 holds for $\Delta = \{\delta_1, \delta_2, \delta_3, \ldots, \delta_n\}$.

The theorem conveys that if

- it holds for all agents $\delta_i \in \Delta$ that they are ignorant of A based on $\forall \delta_i \in \Delta : [C!]I_{\delta_i} A$, and
- that after it has been publicly announced that α knows A, then A is the case, and
- all agents $\delta_i \in \Delta$ know that α knows A, then
- α's knowledge of A will be transferred to every $\delta_i \in \Delta$ provided that
- every $\delta_i \in \Delta$ can mimic α's epistemic behaviour based on theorem 1.

Proof. See Hendricks forthcoming. □

THEOREM 9. $\forall \alpha_i \in \Lambda : I_{\alpha_i} A \wedge ([K_\delta A!](A \wedge K_{\alpha_i} K_\delta A)) \rightarrow [K_\delta A!]K_{\alpha_i} A$ if theorem 2 holds for $\Lambda = \{\alpha_1, \alpha_2, \alpha_3, \ldots, \alpha_n\}$.

- it holds for all agents $\alpha_i \in \Lambda$ that they are ignorant of A based on $\forall \alpha_i \in \Lambda : [C!]I_{\alpha_i} A$, and
- that after it has been publicly announced that δ knows A, then A is the case, and
- all agents $\alpha_i \in \Lambda$ know that δ knows A, then
- δ's knowledge of A will be transferred to every $\alpha_i \in \Lambda$ provided that
- every $\alpha_i \in \Lambda$ can mimic δ's epistemic behaviour based on theorem 2.

Proof. See Hendricks forthcoming. □

Besides the formalization of pluralistic ignorance, the latter theorems demonstrate how some simple versions of pluralistic ignorance may be dispelled using an elementary, and perhaps also less shopworn notion, from epistemic logic which fits well with Floridi's wish for "methodological minimalism" and "artificial agents" with emphasis on computing and information-driven inquiry.

There is so much more of this to be done here, and it is certainly worth our epistemological time—mainstream and formal alike—to consider the interesting social phenomena uncovered by social psychology and behavioral economics. This was a first stab.

References

Bikhchandani, S., D. Hirshleifer, and I. Welch. 1998. "Learning from the Behavior of Others: Conformity, Fads, and Informational Cascades." *Journal of Economic Perspectives* 12:151–70.

Floridi, Luciano. 2008. "Answers to Philosophy of Information: 5 Questions." In *Philosophy of Computing and Information: 5 Questions*, edited by Luciano Floridi, 89–95. New York: Automatic Press/VIP.

Hendricks, Vincent F. 2001. *The Convergence of Scientific Knowledge: A View from the Limit*. Trends in Logic: Studia Logica Library Series. Dordrecht: Springer.

———. 2003 "Active Agents." Φ-*News* 2:5–40. (A revised version of this article appears in a special volume of the *Journal of Logic, Language, and Information* 12, no. 4 [Autumn]: 469–95.)

———. 2004. "Hintikka on Epistemological Axiomatizations." In *Quantifiers, Questions and Quantum Physics: Essays on the Philosophy of Jaakko Hintikka*, edited by Daniel Kolak and John Symons, 3–34. Dordrecht: Kluwer.

———. 2007. *Mainstream and Formal Epistemology*. New York: Cambridge University Press.

———. Forthcoming. *The Agency*. New York: Cambridge University Press.

Hintikka, Jaakko. 1962. *Knowledge and Belief: An Introduction to the Logic of the Two Notions*. Ithaca: Cornell University Press. (Reissued in 2005 by College Publications, London.)

Kelly, Kevin. 1996. *The Logic of Reliable Inquiry*. New York: Oxford University Press.

Roy, Olivier. 2008. *Thinking Before Acting*. Amsterdam: ILLC. (Dissertation Series DS-2008-3.)

van der Hoek, Wiebe, and Arturro Lomuscio. 2004. "A Logic for Ignorance." *Electronic Notes in Theoretical Computer Science* 85, no. 2 (April): 117–33.

van Ditmarsch, Hans, Wiebe van der Hoek, and Barteld Kooi. 2007. *Dynamic Epistemic Logic*. Synthese Library Series, vol. 337. Dordrecht: Springer.

4

MEETING FLORIDI'S CHALLENGE TO ARTIFICIAL INTELLIGENCE FROM THE KNOWLEDGE-GAME TEST FOR SELF-CONSCIOUSNESS

SELMER BRINGSJORD

1. Introduction

In the course of seeking an answer to the Dretskean (2003) question "How do you know you are not a zombie?" Floridi (2005) issues an ingenious, philosophically rich challenge to artificial intelligence (AI) in the form of an extremely demanding version of the so-called knowledge game (or "wise-man puzzle," or "muddy children puzzle")—one that purportedly ensures that those who pass it are self-conscious. We shall call this test "KG_4"; the significance of the subscript will be clear in due course.

In this chapter, on behalf of (at least the logic-based variety of) AI, I take up Floridi's challenge—which is to say, I try to show that this challenge can in fact be met by AI in the foreseeable future. I'm quite convinced that zombies are logically *and* physically possible, and, indeed, that zombies are precisely what logic-based AI, in the long run, will produce (see, e.g., Bringsjord 1995b); that this possibility is enough to refute the view that human consciousness can be replicated through computation (see, e.g., Bringsjord 1999); that the engineering power of logic-based AI (Bringsjord 2008b) is truly formidable; that *any* behavioral test is within the reach of this form of AI (Bringsjord 1995a); and that AI of any variety ought in fact to be guided by the goal of building artificial agents able to pass tests demanding human-level intelligence (Bringsjord and Schimanski 2003). Therefore, it should be easy enough for the reader to understand that I find Floridi's article to be not only relevant but preternaturally so, and that I'm rather motivated to accept his challenge. Of course, anyone convinced not only of AI's ability to eventually create creatures that *appear* to have minds but also of its ability to produce artificial *minds*, will want to show that Floridi's challenge can be surmounted. One of the remarkable aspects of his article is that it targets *both* "weak" and "strong" AI.[1]

---

[1] Briefly: Weak AI: Standard (= Turing-level) computing machines, perhaps suitably connected by sensors and effectors to the external environment, can eventually be engineered

The plan of my chapter is as follows. In the next section (2) a number of preliminary tasks are completed. For example, I explain the different forms of consciousness relevant to Floridi's knowledge game, and set out the structure of his test-based answer to how-do-you-know-you-are-X questions. In section 3, I review the knowledge game as Floridi sets it out, which includes four increasingly demanding versions. Special attention is paid to the reasoning carried out by agents who pass the third and fourth versions of the game. Then I show in section 4 that Floridi's pessimism about the power of robots and zombies to pass KG_4 is unwarranted, in light of my proof-sketch showing that a robot can deduce the solution to this version of the game. Two objections are then rebutted in section 5, and a brief concluding section (6) wraps up the chapter.

2. Preliminaries

A number of preliminaries must be dealt with before we start in earnest. Let's begin with a characterization of the types of consciousness that are central to Floridi's test.

2.1. Types of Consciousness

Following Floridi, we shall distinguish three types of consciousness: *access consciousness* (abbreviated as *a-consciousness*), *phenomenal consciousness* (*p-consciousness*), and *self-consciousness* (*s-consciousness*).[2] This trio is part of the standard terminological furniture of modern philosophy of mind. For example, Block distinguishes between p-consciousness and a-consciousness. The latter concept is characterized by him as follows: "A state is access-conscious ([a]-conscious) if, in virtue of one's having the state, a representation of its content is (1) inferentially promiscuous, i.e., poised to be used as a premise in reasoning, and (2) poised for [rational] control of action, and (3) poised for rational control of speech" (Block 1995, 231). As I have explained elsewhere (Bringsjord 1997), and as Floridi agrees, it's plausible to regard certain extant, mundane computational artifacts to be bearers of a-consciousness. For example, theorem provers with natural-language generation capability, and certainly sophisticated autonomous robots, would qualify. It follows immediately that a zombie would be a-conscious.

---

to match ($\approx$ *simulate*) the outward behavior of human persons. Strong AI: Standard computing machines, perhaps . . . , can eventually be engineered so as to literally *replicate* the inner mental lives of human persons.

[2] Actually, Floridi speaks of *environmental consciousness* rather than a-consciousness, but the two concepts are equivalent, as Floridi himself avers. Floridi says that an agent is environmentally conscious if it "is able to process information about, and hence to interact with, [its] surroundings, its features and stimuli effectively, under normal circumstances" (2005, 417). In the present chapter I run with "a-consciousness" in view of the fact that in AI and philosophy of AI this is a more familiar term.

And now here is Block's characterization of p-consciousness, which matches what Floridi has in mind: "So how should we point to [p]-consciousness? Well, one way is via rough synonyms. As I said, [p]-consciousness is experience. P-conscious properties are experiential properties. P-conscious states are experiential states, that is, a state is [p]-conscious if it has experiential properties. The totality of the experiential properties of a state are 'what it is like' to have it. Moving from synonyms to examples, we have [p]-conscious states when we see, hear, smell, taste and have pains. P-conscious properties include the experiential properties of sensations, feelings and perceptions, but I would also include thoughts, wants and emotions" (Block 1995, 230). What about s-consciousness? Here is Floridi's description of this concept (where "Ag" stands for any agent): "Ag may be *self-conscious* if Ag has a (second- or higher-order) sense of, or is (introspectively) aware of, Ag's personal identity (including Ag's knowledge that Ag thinks) and (first- or lower-order) perceptual or mental experiences (including Ag's knowledge of what Ag is thinking)" (Floridi 2005). As we can see, none of these three definitions is precise, let alone formal. But that is certainly not Floridi's fault. *No one* has formal accounts of these varieties of consciousness on hand, and we shall thus, of necessity, make do with the descriptions given above.[3]

2.2. *Types of Agents*

I further follow Floridi in partitioning the class of relevant agents into three categories; namely, *human persons* (who enjoy *a-*, *p-*, and *s-consciousness*), robots or *artificial agents* (said by Floridi to be "endowed with interactivity, autonomy and adaptability" [2005, 420]), and *zombies* (who have *a-consciousness* but lack *p-* and *s-consciousness*). Hereafter I refer simply to *persons* as the first class (wanting as I do to leave aside, for example, divine persons), *robots* as the second, and *zombies* as the third. Please note that in AI it's common linguistic practice to regard devices to be bona fide artificial agents or robots even when they are remarkably dim. For example, Russell and Norvig (2002) classify computer programs that do no more than compute elementary number-theoretic functions as artificial agents, and the same lattitudinarian approach holds in AI for the domain of robots as well. In Floridi's scheme, and the present chapter's (which premeditatedly inherits directly from Floridi's), *robots* must be capable of interacting with other agents and the external environment, have autonomy, and be adaptable. For a discussion of a continuum of sophistication for robots and artificial agents directly relevant to the present chapter, see Bringsjord, Noel, and Caporale 2000.

---

[3] Despite the fact that we don't have formal definitions of a-, p-, and s-consciousness, it seems clear that there are some logical relations holding between these concepts. For example, it specifically seems clear that anything that is s-conscious is p-conscious. This is a principle Floridi employs and defends in this paper. I happily affirm the principle.

2.3. The Test-Based Answer to the Question

Floridi takes Dretske's question to be one "that can take as an answer 'a way of knowing that, unlike zombies, we are conscious of things,' that is, how one can *possibly* know that one is a zombie" (Floridi 2005, 419). Understood this way, Floridi maintains that there is a test-based way to answer the question. In fact, Floridi generalizes the situation, and explains that there is a test-based way to answer the question *type*: "How do you know that you are an X?" We read: "Good tests usually are informative, that is, they usually are more than just successful criteria of identification of x as [X], because they examine the very process they are testing precisely while the process is occurring, and so they provide the tested agent with a way of [(c1)] showing that he qualifies as a certain kind of agent, [(c2)] knowing that he is that kind of agent, and [(c3)] answering how he knows that he is that kind of agent, by pointing to the passed test and its [(c1)–(c3)] features" (2005, 420).

Where X is any attribute, we can sum up Floridi's approach via figure 1. In this figure, a test is said to include a stem **S**, a question **Q**, and an environment **E**. (Of course, if we were to specify a full "ontology" of testing, we would need to invoke additional categories; for example, testers. But we are streamlining, without loss of generality, and while to facilitate exposition we shall discuss other categories—we shall speak of testers: the prisoners in Floridi's knowledge game—we shall not explicitly build these categories into our explicit representations or into the reasoning of testees.) The stem refers to information that the tester gives the testee before asking the key question **Q**, and the environment consists of information that the testee can gain by sense perception. For example, in the first of the knowledge-game tests presented by Floridi, the "classic" version of the knowledge game, the tester/guard announces to the testees/prisoners that there are five fezzes of a particular color distribution; this information is part of the stem. The question is simply "What color is your fez?" And the environment for a testee includes the color of the fez atop the heads of two other testees. This color can be readily perceived through vision by each of the prisoners.

In order to seek an answer to Dretske's question, we have only to set $X = not\ a\ zombie$ in the schema of figure 1. As we shall see, Floridi believes he has found assignments to **S**, **Q**, and **E**, in the test KG_4, that provide an answer to Dretske's question.

3. The Knowledge-Game Quartet

Floridi (2005) considers a continuum of four versions of the knowledge game, which we shall refer to as KG_1/KG'_1, KG_2, KG_3, and KG_4. (The only difference between KG_1 and KG'_1 is that in the former, each prisoner answers the question separately, whereas in the latter the trio

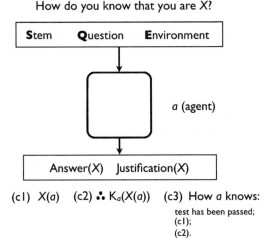

Figure 1. Schematic overview of test-based answer to question type

answers simultaneously as a multiagent system.) In addition, we shall give each version an informal mnemonic label to help us remember something distinctive about a particular version. As you will recall, the last version of the knowledge game, KG_4, is the test for self-consciousness that supposedly separates zombies from persons, and of course supposedly agents from persons as well. This claimed separation is conveyed by table 1, which also expresses the rest of Floridi's claims with respect to whether agents of the three aforementioned types: can pass now ($\sqrt{}$), fail now but possibly pass in the future (?), fail now but will pass in the future (G), or are forever doomed to fail ($\times$).

Table 2 expresses my position, which as the reader can see is rather more optimistic than Floridi's from the perspective of AI.

I shall now briefly review each member of Floridi's continuum.

3.1. The "Classic" Version (KG_1/KG'_1)

The first test in the continuum, KG_1/KG'_1, in many ways serves as a time-honored portal to logic-based formalisms and techniques in AI (see, e.g., Fagin et al. 2004, Arkoudas and Bringsjord 2005, Genesereth and Nilsson 1987), and Floridi sums it up as follows:

A guard challenges three prisoners, A, B, and C. He shows them five fezzes, three red and two blue, blindfolds them and makes each of them wear a red fez, thus minimising the amount of information provided. He then hides the remaining fezzes from sight. When the blindfolds are removed, each prisoner can see only the other prisoners' fezzes. At this point, the guard says: "If you

TABLE 1. Four versions of the knowledge game (Floridi)

Version	Label	Agent Type		
		robots	zombies	Persons
KG_1	"classic"	√	√	√
KG_2	"boots"	√	√	√
KG_3	"deafening"	?	√	√
KG_4	"self-consciousness"	×	×	√

TABLE 2. Four versions of the knowledge game (Bringsjord)

Version	Label	Agent Type		
		robots	zombies	persons
KG_1	"classic"	√	√	√
KG_2	"boots"	√	√	√
KG_3	"deafening"	G	√	√
KG_4	"self-consciousness"	G	G	√

can tell me the colour of your fez you will be free. But if you make a mistake or cheat you will be executed."

The guard interrogates A first. A checks B's and C's fezzes and declares that he does not know the colour of his fez. The guard then asks B. B has heard A, checks A's and C's fezzes, but he too must admit he does not know. Finally, the guard asks C. C has heard both A and B and immediately answers: "My fez is red." (2005, 422)

As astute readers will immediately appreciate, C is quite right, and is therefore released. Readers are expected not only to be able to grasp that C is correct (that is, to grasp that C's fez is red) but also to be able to prove that C's fez is red (using what C knows). For machine-generated and machine-checked proofs that support C's response, see Arkoudas and Bringsjord 2005.

3.2. The "Boots" Version (KG_2)

In KG_2, five pairs of boots are used instead of the fez quintet; two pairs are ordinary, but three are "torturing instruments that crush the feet" (Floridi 2005, 426). The question to the contestants here, of course, is whether one has donned the hurtless variety or the crushing kind. The answer must be given synchronically by the trio.

Floridi is quite right that each type of agent can pass this test with flying colors, and indeed it takes only a modicum of familiarity with the current state of robotics, combined with but a touch of imagination, to grasp that KG_2 could really and truly be passed by today's non-p-conscious and non-s-conscious robots, armed as they are with standard, purely mechanical sensors of various kinds. Therefore, as Floridi correctly asserts, "[b]ootstrapping states are useless for discriminating between humans and zombies" (2005, 426).

3.3. The "Deafening" Version (KG₃)

What distinguishes Floridi's "deafening" version of the knowledge game is that the question **Q** in this case is *self-answering*; such a question is one "that answers itself if one knows how to interpret it" (Floridi 2005, 428). From the perspective of AI (for the basic formal scheme see, e.g., Sun and Bringsjord 2009), this means that a self-answering question Q_{SA}, once parsed by an artificial agent or robot, delivers knowledge φ_Q which, when combined with prior knowledge Φ_p possessed by this agent, allows the agent to infer the correct answer.[4] As a first example, Floridi supplies (**Q4E**) "How many were the four evangelists?"[5]

While Floridi's pessimism about AI's ability to produce s-conscious and p-conscious robots (or zombies), for the purposes of the present dialectic, is to rest upon his KG_4, it is worthwhile to note that this pessimism first surfaces in his paper in connection with KG_4's predecessor, KG_3: Floridi claims that "*current* and *foreseeable* artificial agents [= robots] as we know them cannot answer self-answering questions, either in a stand-alone or in a multiagent setting" (2005, 431). He is certainly right about current robots; he is probably wrong about foreseeable ones.

To see this, consider that, from the standpoint of logic-based AI, engineering a robot that understands and correctly answers (and justifies that answer) some self-answering questions seems surprisingly straightfor-

---

[4] What I say here may strike some alert readers as odd. They may ask: "Don't all questions get answered on the basis of both background knowledge and knowledge (however small it may be) by the question itself? What then distinguishes a *self-answering* question?" A fully satisfying reply would require more space than I have here, but I volunteer that a self-answering question is marked by the fact that answering it correctly can be done without moving outside the bounds of the *a priori* and analytic—as is borne out in the example I very soon give (i.e., **Q3**).

[5] This is actually a rather interesting specimen, because it has a close non-self-answering relative that is effortlessly correctly answered on the basis of *only* standard prior knowledge: "How many were the evangelists?" Or, a less awkward version: "How many evangelists were there?" (There are some unaware of the fact that the quartet in question corresponds to the traditional authors of the four gospels. I don't mean to imply that the background knowledge here is had by everyone. And there are other complications I leave aside, such as that in some heterodox frameworks the writers of the *apocryphal* gospels count as evangelists.)

ward, when you think about some of these questions a bit. For example, consider the self-answering question **Q3**: "What is the cardinality of the set composed of three elements?" Clearly, this question conveys declarative information; specifically, declarative information that captures the nature of the set in question. This information can be expressed in standard first-order logic, following the customary language of set theory; for example:

$$\exists y \exists x_1 \exists x_2 \exists x_3 [a = y \wedge x_1 \in y \wedge x_2 \in y \wedge x_3 \in y \wedge x_1 \neq x_2 \wedge x_2 \neq x_3 \wedge x_1 \neq x_3]$$

If we let this formula be denoted by φ, then a robot seeking to answer **Q3** would be seeking to verify

$$\Phi_{Q3} \vdash \exists n (card(a) = n \wedge \varphi)$$

and this proof can indeed be found by automatic theorem provers armed with the standard machinery of set theory underlying the cardinality of finite sets. Such a proof is elementary, and is found quickly by novices taking their first course in axiomatic set theory.[6] The upshot of this example is that even *current* logic-based AI is able to handle some self-answering questions.

Notice, though, that I say "some" self-answering questions. There is indeed a currently insurmountable obstacle facing logic-based AI that is related, at least marginally, to self-answering questions: it is simply that current AI, and indeed even *foreseeable* AI, is undeniably flat-out impotent in the face of *any arbitrary* natural-language question—whether or not that question is self-answering. To be a bit more precise: Take an artificial agent or robot, stipulate boldly that it's the absolute best that AI can muster today; or bolder still, imagine the best such being that can be mustered by learned extrapolation into the future from where AI is today. Let's dub this wondrous robot "*R*." Now imagine a test that is radically streamlined relative to Floridi's elaborate KG_i, namely, a test in which the question to *R* is just a single fact-finding query; **Q***, let's say. For example: "Is the Vatican south of a tall, largely open-air metal tower located near a river that was home to a famous siege perpetrated by Selmer Bringsjord's violent ancestors?" Of course, the test isn't made any easier if **Q*** happens to be self-answering; the following, for instance, would doubtless stump our *R* as well: "Ceteris paribus, can a superior extemporaneous human debater raised in the United Kingdom learn to play legal chess in an afternoon, if that session is her first exposure to the game?"[7]

---

[6] For sample proofs of this type, quite elementary, see Suppes 1972.

[7] We can of course debate what is and isn't a self-answering question. After all, no identity conditions for such questions have been supplied (by anyone, as far as I can tell; Floridi points out parenthetically that such questions have received surprisingly little attention in logic). The question here, by Floridi's lights, may not be self-answering, but

It should be noted that both Floridi and I refrain from claiming that no robot will *ever* be able to answer arbitrary fact-finding questions. This can be seen by looking at tables 1 and 2, where for KG_3 it will be seen that in Floridi's case he admits that this version of the game may be passed by a future robot ("?"), and in my own case there is the claim that this test is going to be passed in the future ("G").

It should also be noted that Floridi specifically classifies self-answering questions like **Q3** as "internally, semantically self-answering" questions, while the question in the case of the "boots" game is, as he says, "self-answering in a more complex way, for the answer is *counterfactually embedded* in [**Q**] and it is so somewhat 'indexically' since, under different circumstances, the question or the questioning would give nothing away" (2005, 428). The counterfactual aspects of Floridi's depiction of A's reasoning in KG_3 are clearly present; here is that depiction of the reasoning needed to crack KG_3 (where the state S is *hearing the guard's question*, the state D is *being deaf*, and Q is the guard's question): "A reasons that if A were in $\neg S$ then A would be in some state D; but if A were in D then A could not have received Q [= **Q**]; but A received Q, so A could receive Q, so A is not in $\neg S$, but A is in either S or $\neg S$, so A is in S" (Floridi 2005, 429). I end this section by pointing out that *if the counterfactual aspects of Floridi's description of A's reasoning in KG_3 are ignored*, it's easy enough to understand that elementary logic-based AI techniques can be used to express and certify this reasoning.[8] Such understanding arrives once one sees that (1) the core reasoning in standard extensional form is simple, and that (2) such reasoning can be easily certified by any of today's decent automated theorem provers. For the first point, simply confirm mentally that

$$\{\forall x(Sx \to Dx), \forall x(Dx \to \neg RecQx), RecQa, \forall x(Sx \lor (Sx))\} \vdash Sa$$

and then, for the second point, observe that in figure 2, using the Slate system (Bringsjord et al. 2008), the "decounterfactualized" reasoning has been certified by one of the automated theorem provers included in Slate (viz., SNARK; see Stickel et al. 1994).

Of course, it may be asked: "What right have you, though, to decounterfactualize the reasoning?" Since, as I have noted from the outset of this chapter, KG_4, not KG_3, is the most serious challenge to AI that Floridi has fashioned, and since the fourth version of the knowledge game, as Floridi sees the situation, *also* requires counterfactual reasoning of those

that it is is provably consistent with the formal elements I introduced as general constraints on such questions in the previous paragraph.

[8] This means that some necessary conditions for logic-based-AI machines passing the test are satisfied. Any such machine, if passing tests like those Floridi presents, must have a sufficiently expressive underlying language and proof theory (or *argument* theory in nondeductive cases; see Bringsjord 2008a), and a means of using these elements in a reasonable amount of time. We shall later discuss what is *sufficient* to pass the relevant class of tests.

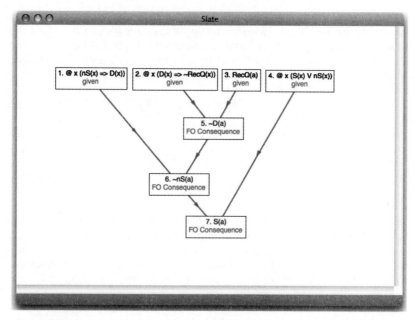

Figure 2. Proof for "cracking" KG_3 machine certified in the Slate system (courtesy of Joshua Taylor)

agents who successfully negotiate it, I will save my answer to this question/objection until I analyze KG_4 (in section 4).

3.4. The "Self-Consciousness" Version (KG_4)

We come now to the positively ingenious "self-consciousness" version of the knowledge game, in which our three embattled prisoners are given not fezzes or boots or beverages but a tablet from a collection of five, three of which are innocuous, while two make those ingesting them completely dumb.

Here is the (once again counterfactual) reasoning that leads A to give the correct answer, in Floridi's words:

> [H]ad I taken the dumbing tablet I would not have been able to report orally my state of ignorance about my dumb/non-dumb state, but I have been and I know that I have been, as I have heard myself speaking and saw the guard reacting to my speaking, but this (my oral report) is possible only if I did not take the dumbing tablet, so now I know that I am in a non-dumb state, hence I know that I have not taken the dumbing tablet, and I know that I know all this, that is, I know that my previous state of ignorance has now been erased, so I can revise my statement and reply, correctly, in which state I am, which is a

state of not having taken the dumbing tablet, of knowing that I haven't, and—
by going through this whole process and passing the test—of knowing how I
know both that I haven't and that I know that I haven't.

(Floridi 2005, 432–33)

4. AI, Contra Floridi, Can Handle KG$_4$

I have argued elsewhere that a grant today of one billion U.S. dollars
would be insufficient to allow a first-rate (for that matter, the world's
preeminent) robotics R&D group to produce, even in an exceedingly long
project, a p-conscious robot (Bringsjord 2007). In fact, I argue that such a
group, however well financed, would not even know where to *start*.
Things are radically different in the case of KG$_4$. In this case, it can be
demonstrated that foreseeable AI can produce an artifact capable of
passing. The demonstration consists in showing that such an artifact can,
now, if sufficient time and energy is expended to carry out the engineering, apparently be produced. While I don't have enough space here to
supply the demonstration in the form of a proof, I *do* have sufficient space
to provide a proof-*sketch*, which I give below. This reasoning, expressed
in significant part in the *cognitive event calculus*, **CEC** for short (for
formal details, see Arkoudas and Bringsjord 2009), should be more than
detailed enough to justify the assertion that *foreseeable* AI will be up to
the task.[9]

Before I present the proof-sketch, a brief overview of the language of
CEC is in order. The reason is that some readers may be unfamiliar with
multisorted logic (MSL), and with some of the core concepts in the event-
calculus approach to reasoning about time and change.

In standard first-order logic (FOL), as is well known, quantification in
any formal theory, or in any knowledge base for some problem or
application, is over a so-called *domain*, which is simply a set. For example,
suppose that we have a domain D composed of all the people in the
classroom of some introductory logic course on some particular day.
There are students in D, as well as teaching assistants, and there is a
professor (a) at the front of the room in question. Given this setup, to say
in standard FOL that every student likes every teaching assistant who
likes the professor, we might write

$$\lambda: = \forall x \forall y ((Sx \wedge Sy \wedge Lya) \rightarrow Lxy)$$

But this is somewhat cumbersome, and—for reasons that certainly
needn't be given here—inefficient when it comes time for a machine to
reason deductively over such information. MSL allows us to partition D so

---

[9] If I were given a grant of sufficient size, I'm confident that with help from colleagues in
my laboratory I could produce a working KG$_4$-passing robot within one year.

that it contains the following *sorts*: Students, TeachingAssistants, and Professors. If we then correspondingly partition our supply of variables, and use obvious notation to indicate doing so, we can supplant λ with

$$\lambda' := \forall s \forall t (Lta \rightarrow Lst)$$

Now, **CEC** is a sorted calculus; it includes the following sorts:

Object Agent ActionType Action Event Moment Boolean Fluent

where Action is a subsort of Event. It's important to know that Boolean is simply the set composed of true and false. In addition, put intuitively, a fluent is a state of affairs that, through time, can shift between holding and not. Fluents have long been used in logic-based AI.

How does one define well-formed formulas (wffs) in this approach? The grammar is a straightforward close relative of those used for FOL; a few remarks will suffice. **CEC** has a number of key function symbols. Here are four defined:

- *holds*: Fluent × Moment → Boolean
- *happens*: Event × Moment → Boolean
- *clipped*: Moment × Fluent × Moment → Boolean
- *initiates*: Event × Fluent × Moment → Boolean

Read informally, the first bullet says that *holds* is a function that takes a fluent and a time point and returns a truth-value giving whether or not the fluent holds at this time point. The second and fourth should be self-explanatory. The third, understood intuitively, conveys that the function *clipped* takes a time point, a fluent, and another time point, and is true when the fluent is terminated between these two times.

We are now in position to articulate the proof-sketch. Here goes:

––––––––––

Theorem. I, A, didn't ingest a tablet of the dumbing variety.

Proof-sketch. We begin by noting that KG_4 pivots around five time points, which we shall make privileged constants in the following manner:

- t_1 (*apprise*): This is the time point at which the prisoners are apprised of the situation; that is, the time at which they learn of the five tablets (and the two kinds therein, partitioned, recall, 3–2), and so on. In short, a good portion of the knowledge acquired from the environment **E** is here perceived by A. (I shall use "a" rather than "A" because constants are traditionally lowercase characters, and at any rate they are in **CEC**; see Arkoudas and Bringsjord 2009.) I assume that this information consists of formulae in the set Φ.

- t_2 (*ingest*): The pills are ingested by the quartet. *A*, of course, consumes a nondumbing tablet.
- t_3 (*inquire*): The guard inquires as to which tablet has been taken.
- t_4 (*speak1*): *A* responds with "Heaven knows!"
- t_5 (*speak2*): *A* says: "The nondumbing variety!" (Alternatively, *A* responds with "No!" in response to the question "Did you receive a dumbing tablet?")

We observe that *a* perceives the first part of the information he will soon enough use to pass the test:

(1) **P** (*a*, Φ, *apprise*)

Here I make use of the **P** operator for perceiving. This is a slight adaptation of the **S** operator in *CEC*, which represents seeing. Since we are dealing with a fair test, we know that optical and auditory illusions can be safely ignored, and hence have a rule (which I don't bother to reproduce verbatim; this is a proof-*sketch*, after all) allowing agents to infer that which they directly perceive.

It follows immediately from (1) and the rule known as DR_4 in *CEC* (again, for full formal details, see Arkoudas and Bringsjord 2009), viz.,

$$\mathbf{P}(agent, \phi) \Rightarrow \mathbf{K}(agent, \phi)$$

that[10]

(2) **K** (*a*, Φ, *apprise*)

Next, note specifically that *a* knows that if he takes the dumbing tablet at *ingest*, he can't report orally his state of ignorance at any subsequent time, unless the effects are—to use the language of the event calculus— "clipped." This result corresponds to counterfactual knowledge, in Floridi's informal version of *A*'s reasoning; keep this point in mind, for it will be quite relevant shortly (in section 5.2). I can prove this (lemma) now. First, *a* has the following knowledge on the basis of Φ:

(3) **K** (*a*, (*initiates* (*action* (*a*, *eats*(*p*)), *dumb$_a$*, *ingest*)))

Next, note that *a* knows that if no clipping occurs, then he will remain dumb. More precisely, he can deduce that if the left-hand side of the biconditional inside the third axiom shown in footnote 12 (i.e., A3) is negated, then no clipping of the poison occurs—in which case he is dumb at all time points later than *ingest*. The deduction here is easy, given the fact that what is known is true,[11] combined with the standard axioms of

---

[10] To ease and accelerate exposition, I overload the **K** operator. In *CEC* proper, this operator ranges over the agent in question, and some proposition *P*. Here, I compress declarative information by allowing the operator to range over *sets* of propositions, and to have a third argument (a time point).

[11] The rule in Arkoudas and Bringsjord 2009 is R_4 and is this: $\mathbf{K}(agent, \phi) \Rightarrow \phi$

the event calculus, which I assume to be common knowledge among the prisoners.[12]

But of course we are not home yet: our agent a must deduce that he did *not* receive a tablet of the dumbing variety. Thankfully, the deduction is easy.

The high-level structure of the deduction conforms to indirect proof: the assumption that agent a *did* receive a dumbing tablet leads to a contradiction, from which we infer by reductio ad absurdum that our assumption is in error, and hence the answer from a is a negative one.

We have already seen that if a did receive a dumbing tablet, then at all time points he cannot speak; hence he cannot speak at the particular time point *speak1*. Suppose for the sake of argument, then, that a did receive a dumbing tablet. Then by our lemma he cannot speak at *speak1*. But a perceives that he *does* speak at this time point. Hence he knows that he speaks at this time point. Hence he does in fact speak at this time point. Therefore a contradiction is produced. By reductio, a did not receive a dumbing tablet. **QED**

———

Hypersedulous readers are encouraged to flesh out my reasoning so as to produce a step-by-step proof. For the benefit of such folks (and to any of them yet to obtain the Ph.D. who produce the proof: please contact me about potential graduate study, immediately), I divulge that two unspoken axioms are needed for the detailed version:

- "All agents know that they know of the events they intentionally bring about themselves." Formalized:

  $\mathbf{C}$ ($\forall$ a, d, t (*happens* (*action* (a, d), t)) $\rightarrow$ $\mathbf{K}$ (a,*happens* (*action* (a, d), t))))

- "All agents know that if an agent believes that a certain fluent f holds at t and that agent doesn't believe that f has been clipped between t and t', then that agent will believe that f holds at t'." Formalized:

  $\mathbf{C}$ ($\forall$ a, f, t, t' (($\mathbf{B}$ (a, *holds*(f, t)) $\wedge$ $\mathbf{B}$ (a, $t < t'$) $\wedge$ $\neg$ $\mathbf{B}$ (a, *clipped*(t, f, t'))) $\rightarrow$ $\mathbf{B}$ (a, *holds*(f, t')))

---

[12] Hence we have, where $\mathbf{C}$ is the common knowledge operator:

- A1: $\mathbf{C}$ ($\forall$ f, t (*initially*(f) $\wedge$ $\neg$ *clipped*($0, f, t$) $\rightarrow$ *holds*(f, t)))
- A2: $\mathbf{C}$ ($\forall$ e, f, t_1, t_2 ((*happens*(e, t_1) $\wedge$ *initiates*(e, f, t_1) $\wedge$ $t_1 < t_2$ $\wedge$ $\neg$ *clipped*(t_1, f, t_2)) $\rightarrow$ *holds*(f, t_2))
- A3: $\mathbf{C}$ ($\forall$ t_1, f, t_2 (*clipped*(t_1, f, t_2) $\leftrightarrow$ ($\exists$ e, t (*happens*(e, t) $\wedge$ $t_1 < t < t_2$ $\wedge$ *terminates*(e, f, t))))))

Note that since the common knowledge operator $\mathbf{C}$ is applied to each axiom, it can be instantly deduced that all prisoners/agents know these axioms, from which it follows by R_4 that these axioms are true in this context.

5. Objections

5.1. "But A's reasoning is first-person reasoning"

The first objection runs as follows: "Your proof sketch, Bringsjord, which is intended to show a fully mechanical version of the reasoning Floridi ascribes to prisoner A, dodges the fact that we are talking here about *self-consciousness*. Notice the use of the first-person pronoun in the reasoning that Floridi presents as an expression of A's. This pronoun is absent in your proof sketch; you use only the constant a, not 'I.' Hence you have failed to capture A's solution."

This objection is easily dispensed with.

First, as a matter of formal logic and the specifics of **CEC**, the fact is that restricted versions of the epistemic version of the modal axiom 4 (which marks the modal system KT4/S4) are active in the present case.[13] I can't discuss the specifics here, but the point is that knowing P essentially implies knowing that one knows P—which is a phenomenon often closely associated with first-person knowledge. (The cognitive event calculus, on the other hand, does *not* allow infinite iteration of knowledge operators. Only three iterated **K**s are permitted in any formula.)

Second, recall the traditional tripartite *de dicto/de re/de se* distinction with respect to kinds of beliefs that has become standard in rigorous epistemology. We can quickly encapsulate the distinction by listing examples (slightly adapted to present purposes) of the three types given by Chisholm (1981, 18):

- *de dicto*: The tallest man believes that the tallest man is wise.
- *de re*: There is an x such that x is identical with the tallest man, and x is believed by y to be wise.
- *de se*: The tallest man believes that he himself is wise.

My view, and the one that underlies the proof-sketch given earlier, is that Frege (1956), Husserl (1970), and Chisholm (1976) are correct that all *de re* and *de se* belief can be reduced to *de dicto* belief, given that persons are associated with individual essences, semantically. Here is how Chisholm summarizes the reduction view in *The First Person*: "Some philosophers—for example, Frege and Husserl—have suggested that each of us has his own idea of himself, his own *Ich-Vorstellung* or individual concept. And some of the things that such philosophers have said suggest the following view: The word 'I,' in the vocabulary of each person who uses it, has for its referent that person himself and has for its sense that person's *Ich-Vorstellung* or individual concept. The difference between my 'I'-propositions and yours would lie in the fact that mine imply my

---

[13] Nonepistemic axiom 4 is: $\Box \varphi \rightarrow \Box \Box \varphi$. A good discussion at the propositional level is provided by Chelles 1980. A good discussion at the quantified level (and note that **CEC** is at this more expressive level) is provided by Hughes and Cresswell 1968.

Ich-Vorstellung and not yours, and that yours imply your *Ich-Vorstellung* and not mine" (1981, 16). We don't need to analyze here the ins and outs of essences. We need not plumb the depths of the question of whether, as a matter of metaphysics, persons have individual essences or haecceities. The point is that whether or not they do, the view that they do suggests a corresponding move in formal logic that serves to help mathematize and mechanize *A*'s reasoning. From the logico-computational viewpoint, the role that individual essences are to play in the production of the above proof sketch is clear: that is the role of allowing, formally, the needed reduction. And Chisholm shows how to carry out the reduction, in chapter 1 of *Person and Object* (1976). The basic trick is perfectly straightforward: *De re* belief is belief that a relevant proposition holds. What proposition? Consider the tallest-man trio of examples given above. Consider, specifically, the situation in which you are the tallest man; you are (as you most assuredly are) wise; and I believe in *de re* fashion of you that you are wise. Then on Chisholm's view I believe a proposition φ which deductively implies that you have both the properties *being the tallest man* and *being wise*. Our φ here is just the proposition that the tallest man is wise.

Of course, there isn't space here to cover the reduction in any detail. Given present purposes, it suffices to note that the reduction requires that each person (and in the case of the machine-generated correlate of *A*'s reasoning as conveyed by Floridi, each computing machine) be associated with an individual essence, in our formal semantics. We can thus say that while *a* is an ordinary constant in the language of the cognitive event calculus, and hence it's entirely possible for *a* to be identical to some other constant (for example, the proper name of prisoner *A*; Alfred, perhaps), *a** is a symbol functioning as a personal pronoun for prisoner *A*. We have then only to amend my proof-sketch by replacing each occurrence of *a* with *a**—and we are done.[14]

---

[14] How would the details look? The simplest thing to do (and I am of course under no obligation to provide a formal semantics that is complicated; all I need is something that gets the engineering job done) is to give a "syntactic" semantics for **CEC** based simply on sets, directly. On this approach, what an agent *b* knows is simply collected into a set (a box) of formulae, suitably indexed with her name. What she knows she knows is simply collected into a box within her box. This approach, which is classically set out in Genesereth and Nilsson 1987, could be appropriated for present purposes without requiring too much imagination, starting with legislation to the effect that every agent *b* has on the semantic side a "haecceity" symbol H_b associated with him, and continuing with the stipulation that first-person beliefs are not only the relevant standard formulae in *b*'s box, but the injection, at the relevant time point, of the string H_b (b) into that box. My intuitive understanding of this string would correspond to what Floridi is quoted as saying above in the underlined key phrase in the quote I gave in section 2.1: <u>Ag's personal identity</u>. Note that Chisholm even specifically says in *Person and Object* (1976) that one's haecceity may consist in the property of being identical to oneself.

5.2. "But A's reasoning is counterfactual reasoning"

The second objection: "You yourself conceded, Bringsjord, that even the reasoning given to solve KG$_3$ was counterfactual in nature; and in fact you agreed that the reasoning in KG$_4$ is—as Floridi claims—counterfactual as well. But your proof-sketch appears to be based solely on the *material* conditional. That conditional is allowed to be within the scope of various epistemic and perceptual operators (e.g., **C**, **K**, **P**, etc.), but this in no way yields a conditional that is counterfactual in nature. Hence it's clear that your proof-sketch fails to point the way to a machine-based version of A's victorious reasoning, as set out by Floridi."

In reply, first note that I didn't say that reasoning that produces a pass in the case of the KG$_4$ test *must* employ counterfactual conditionals. I grant only that the reasoning Floridi offers on behalf of A makes use of natural-language versions of such conditionals.[15] The event calculus, as a matter of mathematical fact, obviates, in certain contexts, the need for counterfactual conditionals. In short, my future KG$_4$-passing machine agent has no need of such conditionals, because their import is expressed by way of the branching histories that the event calculus secures.[16]

My rebuttal can be viewed from a different perspective, namely, that of a judge in the case of KG$_4$. Suppose, in fact, that you are in the role of judge and must decide the fate of A, based on his response (his answer **A** and justification **J**; recall again figure 1). Now suppose that A provides not only "Heaven knows!" and (if **Q** is "Did you ingest a dumbing tablet?") "No!" and a natural-language version of the proof sketch I have supplied. Would not the judge, upon receiving this content from A, declare that the test had been passed? I should think so.

The final part of my reply is simply to note that, from the standpoint of the field that arguably bears most directly on the challenge facing prisoner A, decision theory, object-level counterfactual reasoning is not necessary. To put it starkly, decision theory, even when elaborate and philosophically sophisticated, can be erected in the complete absence of the niceties of conditional logic (see, e.g., Joyce 1999).

---

[15] In this connection, it's worth nothing that the tradition surrounding KG$_1$ and its relatives (e.g., the muddy children puzzle) is one in which formal logic-based modeling need not reflect counterfactuals. See, e.g., Fagin et al. 2004.

[16] Though I can't present any of the details here, even if Floridi insisted that a machine-generated and machine-certified proof corresponding to A's success include not → but the > discussed, e.g., by Nute (1984), the situation could be handled by formalizing the semantics for > in extensional first-order logic, and relying after that on standard automated proving power of the sort that allowed figure 2 to be produced. Such a trick is actually essentially one that parallels the one used to reduce inference in **CEC** to inference in standard first-order logic (for details see Arkoudas and Bringsjord 2009).

6. Conclusion

Floridi succeeds in delivering an inventive, unprecedentedly difficult challenge to logic-based AI; this much I gratefully concede.[17] However, in light of the foregoing, it is seen that this challenge can be met by foreseeable AI. Is there a further variation of the knowledge game beyond the capacity of robots produced by foreseeable AI? Yes, I believe so; and I believe that Floridi may eventually find and disclose such a variation. Is there a variation of the game beyond the capacity of a robot to solve, period? No. The problem is that the knowledge game, as a fair, *empirical* test, is by definition such that there is some finite, observable behavior, β, which the judge, as a perfectly rational agent employing certain principles for decision making, will be correct in judging to be a "passing grade." Since such a β can be generated by a suitably programmed Turing machine (or equivalent) operating over a finite amount time without contravening the laws of logic or even physics, it can immediately be deduced that there exists (in so-called mathematical space) a robot that passes with flying colors. To put the point another way, there is every reason to think that as AI marches on into a future beyond our children, and our children's children, and indeed into time centuries hence, leaving us at best a distant memory, the universe will be populated by computational creatures whose external behavior includes anything and everything within our own repertoire. (It is perhaps ironic, but certainly true, that none other than Floridi himself is among the very few on our planet who have professionally and prudently—and in my opinion prophetically—pondered a future in which the world is saturated with information; see, e.g., Floridi 2007.) These creatures may nonetheless lack s-consciousness and p-consciousness. We might know that they lack these attributes, but only via extensive reasoning showing that s-consciousness or p-consciousness is more than computation (see, e.g., the arguments in Bringsjord and Zenzen 2003). But only God would know *a priori*, because his test would be direct and nonempirical: he would know whether these beings are s- and p-conscious not by following the recipe of figure 1 but by considering whether or not such consciousness is present, end of story—analogous to our ability to settle, quite independent of empirical testing, whether or not, say, 83 is a prime number.

---

[17] I can't rationally declare that his challenge is supremely difficult for *other* forms of AI—say, for "low-level" robotics as opposed to logic-based robotics, as classically defined by Levesque and Lakemeyer (2007). The reason includes, for instance, that the knowledge game simply doesn't require intricate, nonsymbolic, "perception-and-action" engineering. For example, it doesn't demand that robots display human-level physical manipulation.

Acknowledgments

I'm greatly indebted to Luciano Floridi—not only for an inspiring, ever-expanding (Zeno-like) body of written work (a microscopic part of which I engage here) but also for many stimulating conversations about matters at the intersection of philosophy of mind and computation/information, including, specifically, the matters I discuss in the present chapter. Massive thanks are due to Konstantine Arkoudas for work in logic-based AI exploited here. I'm grateful as well to Micah Clark, an anonymous referee, and Patrick Allo for many insightful comments.

References

Arkoudas, K., and S. Bringsjord. 2005. "Metareasoning for Multi-Agent Epistemic Logics." In *Fifth International Conference on Computational Logic in Multi-Agent Systems (CLIMA 2004)*, vol. 3487 of Lecture Notes in Articial Intelligence (LNAI), 111–25. New York: Springer. URL: http://kryten.mm.rpi.edu/arkoudas.bringsjord.clima.crc.pdf.

———. 2009. "Propositional Attitudes and Causation." *International Journal of Software and Informatics* 3, no. 1:47–65. URL: http://kryten.mm.rpi.edu/PRICAI w sequentcalc 041709.pdf.

Block, N. 1995. "On a Confusion About a Function of Consciousness." *Behavioral and Brain Sciences* 18:227–47.

Bringsjord, S. 1995a. "Could, How Could We Tell if, and Why Should—Androids Have Inner Lives?" In *Android Epistemology*, edited by K. Ford, C. Glymour, and P. Hayes, 93–122. Cambridge, Mass.: MIT Press.

———. 1995b. "In Defense of Impenetrable Zombies." *Journal of Consciousness Studies* 2, no. 4:348–51.

———. 1997. "Consciousness by the Lights of Logic and Common Sense." *Behavioral and Brain Sciences* 20, no. 1:227–47.

———. 1999. "The Zombie Attack on the Computational Conception of Mind." *Philosophy and Phenomenological Research* 59, no. 1:41–69.

———. 2007. "Offer: One Billion Dollars for a Conscious Robot. If You're Honest, You Must Decline." *Journal of Consciousness Studies* 14, no. 7:28–43. URL: http://kryten.mm.rpi.edu/jcsonebil lion2.pdf.

———. 2008a. "Declarative/Logic-Based Cognitive Modeling." In *The Handbook of Computational Psychology*, edited by R. Sun, 127–69. Cambridge: Cambridge University Press. URL: http://kryten.mm.rpi.edu/sb lccm ab-toc 031607.pdf.

———. 2008b. "The Logicist Manifesto: At Long Last Let Logic-Based AI Become a Field unto Itself." *Journal of Applied Logic* 6, no. 4:502–25. URL: http://kryten.mm.rpi.edu/SB LAI Manifesto 091808.pdf.

Bringsjord, S., R. Noel, and C. Caporale. 2000. "Animals, Zombanimals, and the Total Turing Test: The Essence of Articial Intelligence."

Journal of Logic, Language, and Information 9:397–418. URL: http://kryten.mm.rpi.edu/zombanimals.pdf.

Bringsjord, S., and B. Schimanski. 2003. "What Is Articial Intelligence? Psychometric AI as an Answer." In *Proceedings of the 18th International Joint Conference on Articial Intelligence (IJCAI–03)*, 887–93. San Francisco: Morgan Kaufmann. URL: http://kryten.mm.rpi.edu/scb.bs.pai.ijcai03.pdf.

Bringsjord, S., J. Taylor, A. Shilliday, M. Clark, and K. Arkoudas. 2008. "Slate: An Argument-Centered Intelligent Assistant to Human Reasoners." In *Proceedings of the 8th International Workshop on Computational Models of Natural Argument (CMNA 8), Patras, Greece*, edited by F. Grasso, N. Green, R. Kibble, and C. Reed, 1–10. URL: http://kryten.mm.rpi.edu/Bringsjord etal Slate cmna crc 061708.pdf.

Bringsjord, S., and M. Zenzen. 2003. *Superminds: People Harness Hypercomputation, and More*. Dordrecht: Kluwer Academic.

Chellas, B. F. 1980. *Modal Logic: An Introduction*. Cambridge: Cambridge University Press.

Chisholm, R. 1976. *Person and Object: A Metaphysical Study*. London: George Allen and Unwin.

———. 1981. *The First Person: An Essay on Reference and Intentionality*. Minneapolis: University of Minnesota Press.

Dretske, F. 2003. "How Do You Know You Are Not a Zombie?" In *Privileged Access and First-Person Authority*, edited by B. Gertler, 1–13. Burlington: Ashgate.

Fagin, R., J. Halpern, Y. Moses, and M. Vardi. 2004. *Reasoning About Knowledge*. Cambridge, Mass.: MIT Press.

Floridi, L. 2005. "Consciousness, Agents and the Knowledge Game." *Minds and Machines* 15, nos. 3–4:415–44. URL: http://www.philosophyonformation.net/publications/pdf/caatkg.pdf.

———. 2007. "A Look into the Future Impact of ICT on Our Lives." *Information Society* 23, no. 1:59–64.

Frege, G. 1956. "The Thought: A Logical Inquiry." *Mind* LXV 289–311.

Genesereth, M., and N. Nilsson. 1987. *Logical Foundations of Articial Intelligence*. Los Altos, Calif.: Morgan Kaufmann.

Hughes, G., and M. Cresswell. 1968. *An Introduction to Modal Logic*. London: Methuen.

Husserl, E. 1970. *Logical Investigations*. London: Routledge and Kegan Paul.

Joyce, J. 1999. *The Foundations of Causal Decision Theory*. Cambridge: Cambridge University Press.

Levesque, H., and G. Lakemeyer. 2007. "Cognitive Robotics." In *Handbook of Knowledge Representation*, edited by F. van Harmelen, V. Lifschitz, and B. Porter, 869–86. Amsterdam: Elsevier.

Nute, D. 1984. "Conditional Logic." In *Handbook of Philosophical Logic, vol. 2: Extensions of Classical Logic*, edited by D. Gabbay and F. Guenthner, 387–439. Dordrecht: D. Reidel.

Russell, S., and P. Norvig. 2002. *Articial Intelligence: A Modern Approach*. Upper Saddle River, N.J.: Prentice Hall.

Stickel, M., R. Waldinger, M. Lowry, T. Pressburger, and I. Underwood. 1994. "Deductive Composition of Astronomical Software from Subroutine Libraries." In *Proceedings of the Twelfth International Conference on Automated Deduction (CADE–12), Nancy, France*, 341–55. SNARK can be obtained at the URL provided here. URL: http://www.ai.sri.com/∼ stickel/snark.html

Sun, R., and S. Bringsjord. 2009. "Cognitive Systems and Cognitive Architectures." In *The Wiley Encyclopedia of Computer Science and Engineering*, edited by B. W. Wah, 1: 420–28. New York: Wiley. URL: http://kryten.mm.rpi.edu/rs sb wileyency pp.pdf.

Suppes, P. 1972. *Axiomatic Set Theory*. New York: Dover.

5

INFORMATION WITHOUT TRUTH

ANDREA SCARANTINO AND GUALTIERO PICCININI

1. Information and the Veridicality Thesis

According to the Veridicality Thesis, information requires truth. More precisely, our main focus is the thesis that if system A has information about the obtaining of p, then p.[1] On this view, smoke carries information about there being a fire only if there is a fire, the proposition that the earth has two moons carries information about the earth having two moons only if the earth has two moons, and so on. We reject this Veridicality Thesis. We will argue that the main notions of information used in cognitive science and computer science allow A to have information about the obtaining of p even when p is false.

Our rejection of the Veridicality Thesis is part of our effort to explicate the views that (i) cognition involves information processing and (ii) computing systems process information. Disentangling various threads in these views, which are ubiquitous in cognitive science and computer science, is part of a broader investigation into the foundations of cognitive science (Piccinini and Scarantino forthcoming). The take-home message of this chapter is that a healthy pluralism in the theory of information leads to positing kinds of information that are at odds with the Veridicality Thesis.

2. Information as a Mongrel Concept

Accounts of information can be motivated by either of two assumptions. The first is that the disparate uses of the term "information" in ordinary language and science can be captured by a unique, all-encompassing notion of information. The second is that different uses of "information" reveal a plurality of notions of information, each in need of a separate

---

[1] By "system A has information about the obtaining of p" we mean that information about the obtaining of some p is available to system A. We are neutral on whether A is in a position to interpret, and make use of, the information available to it. This formulation of the Veridicality Thesis is provisional: we will later distinguish between two versions of the Veridicality Thesis, one for "natural" and the other for "nonnatural" information.

account. We endorse this second assumption: information is a mongrel concept in need of disambiguation.

The allure of the Veridicality Thesis may be due at least in part to a lack of distinction between two notions of *semantic* information. Generally speaking, semantic information has to do with the semantic content, or meaning, of signals.[2] There are at least two important kinds of *meaning*, namely, natural and nonnatural meaning (Grice 1957). Natural meaning is exemplified by a sentence such as "Those spots mean measles," which is true—Grice claimed—just in case the patient has measles. Nonnatural meaning is exemplified by a sentence such as "Those three rings on the bell (of the bus) mean that the bus is full" (1957, 85), which is true even if the bus is not full.

In an earlier essay, we extended the distinction between Grice's two types of *meaning* to a distinction between two types of *semantic information*: *natural information* and *nonnatural information* (Piccinini and Scarantino forthcoming). Spots carry natural information about measles by virtue of a reliable physical correlation between measles and spots.[3] By contrast, the three rings on the bell of the bus carry nonnatural information about the bus being full by virtue of a convention.[4] The two notions of information are importantly different, so the discussion on whether information entails truth needs to reflect both. Although other authors are not always clear on the distinction between natural and nonnatural information, natural information is close to the notion most often discussed by philosophers (Dretske 1981, 1988; Fodor 1990; Millikan 2004; Cohen and Meskin 2006, 2008; see also Stampe 1975, 1977).

At this point, a critic may interject that the notion of nonnatural information is problematic, because it presupposes intentional content (say, the content *that the bus is full*). Because of this, nonnatural information cannot be used to naturalize intentionality, which is one of the central projects that have attracted philosophers to information in the first place (Dretske 1981, 1988; Millikan 2004). Hence, several authors working within this tradition have been adamant in wanting to distinguish information, understood along the lines of Grice's natural meaning, from meaning, understood along the lines of Grice's nonnatural meaning.

We agree that nonnatural information is in need of naturalistic explication just as much as intentional content. "To carry nonnatural information about" and "to represent" are (for present purposes) synonymous expres-

---

[2] Notions of semantic information contrast with notions of nonsemantic information such as the one developed by Shannon (1948) for engineering purposes. What interested Shannon was not the meaning of signals but rather their probability of being selected.

[3] The notion of *natural information* is similar to what Floridi (2008) calls *environmental information*. Unlike Floridi, we follow Dretske (1981) in considering *natural/environmental information* a semantic notion, despite the absence of an intelligent producer of the signals.

[4] This is not to say that conventions are the only possible source of nonnatural meaning. For further discussion, see Grice 1957.

sions, and representation is precisely what a naturalistic theory of intentionality aims to explicate. But our present goal is not to naturalize intentionality. Rather, it is to understand the central role played by information in the sciences of mind and in computer science.

If cognitive scientists and computer scientists used "information" only to mean natural information, we would happily follow the tradition, and speak of information exclusively in its natural sense. The problem is that the notion of information as used in the special sciences often presupposes intentional content.

Instead of distinguishing sharply between information and meaning, we distinguish between natural information, understood (roughly) along the lines of Grice's natural meaning,[5] and nonnatural information, understood along the lines of Grice's nonnatural meaning. We lose no conceptual distinction, while we gain an accurate description of how the concept of information is used in the sciences of mind and in computer science. We take this to be a good bargain.

One payoff of our taxonomy is that we explicitly distinguish between a notion of information that can help naturalize intentionality (natural information) and a notion of information that itself needs to be naturalized (nonnatural information). Another payoff is that we minimize the risk of cross-purpose talk, which often occurs when different theorists speak of information while meaning importantly different things by it. We hope that, once the options are clear, theorists will make explicit which notion of information they are committed to.

We will now argue against the Veridicality Thesis, discussing the cases of natural information and nonnatural information in turn.

3. Natural Information Without Truth

Can spots carry natural information about measles in the absence of measles? Can smoke carry natural information about fire when there is no fire? Can a ringing doorbell carry natural information about someone being at the door even if no one is? According to Grice (1957), the answer to these questions would seem to be no. These are all cases of natural meaning, which is supposed to differ from nonnatural meaning precisely because it is truth entailing.

We depart from this standard picture by holding that the transmission of natural information entails nothing more than the truth of a probabilistic claim. The core idea of the probabilistic theory of information one of us has developed is that signals carry natural information by changing the probability of what they are about (Scarantino unpublished). On this view, spots carry natural information about measles not

---

[5] One important difference is that, unlike Grice, we reject the Veridicality Thesis for natural information, at least in the formulation we consider in section 3.

because all and only patients with spots have measles but because patients with spots are more likely to have measles than patients without spots.[6]

A corollary of this view is that token signals can carry natural information about events that fail to obtain. So far we have formulated the Veridicality Thesis as the thesis that if system A has information about the obtaining of p, then p. This formulation obscures the distinction between natural information that o is *probably* G (henceforth, *probabilistic* natural information) and natural information that o is G (henceforth, *all-or-nothing* natural information). As we argue below, the distinction between the two cases is crucial to a full understanding of natural information.

To frame the issue of natural information correctly, the Veridicality Thesis should be reformulated. Following Dretske (1981), we take the relata of the natural information relation to be *events*, understood as property exemplifications at a time (cf. Kim 1976). The Veridicality Thesis for Natural Information (VT_N) is the following:

> (VT_N) If a signal s being F carries natural information about an object o being G, then o is G.

A corollary of VT_N is that if an agent A has natural information about an object o being in state G when she receives a signal s in state F, then o is G. By contrast, here is the Probability Raising Thesis for Natural Information (PRT_N):

> (PRT_N) If a signal s being F carries natural information about an object o being G, then $P(o$ is $G|s$ is $F) > P(o$ is $G| \sim (s$ is $F))$.[7]

On this view, a signal s in state F can carry natural information about an object o being G simply by raising the probability that o is G, whether or not o is in fact G. We reject VT_N in favor of PRT_N.

Several contemporary theories of natural information are committed to VT_N. For instance, Cohen and Meskin argue that s being F carries information about o being G if and only if the counterfactual conditional "if o were not G, then s would not have been F" is nonvacuously true (2006, 335).[8] In

---

[6] In Piccinini and Scarantino forthcoming, we describe natural information as truth entailing, suggesting that our distinction between natural and nonnatural information parallels Grice's (1957) distinction between natural and nonnatural meaning. It should now be clear that our distinction differs from Grice's in one important respect. On our view, there is nothing objectionable in holding that "those spots carry natural information about measles, but he doesn't have measles," provided measles is more likely given those spots than in the absence of those spots.

[7] One of us has argued that signal s being F can also carry (negative) natural information about o being G by lowering the probability that o is G (Scarantino unpublished). We disregard this complication in what follows.

[8] In the absence of the nonvacuity proviso, anything would carry information about o being G whenever it is necessary that o is G. In such a case, "if o were not G then s would not be F" would have an impossible antecedent, and counterpossibles are commonly assumed to be vacuously true (Lewis 1973).

other words, in order for smoke to carry natural information about fire, it must be that, had fire not occurred, smoke would not have occurred. From this it follows that if smoke carries natural information about the obtaining of a fire, a fire has to obtain. Thus, Cohen and Meskin's (2006) theory entails VT_N.[9]

The notion of natural information presupposed by VT_N is all-or-nothing: either a signal carries *the* natural information *that o* is *G*, or it carries *no* natural information about *o* being *G*. Theorists of natural information have focused mainly on all-or-nothing natural information. This is due to their interest in a theory of information as a chapter in a theory of knowledge. For instance, Dretske argues that "[i]nformation is what is capable of yielding knowledge, and since knowledge requires truth, information requires it also" (1981, 45). This link between information and knowledge is a primary motivation for the idea that information entails truth.

The goal of an information-based account of knowledge has led many to neglect cases of probabilistic natural information that do not yield knowledge, yet fully qualify as information. In other words, the true thesis that information is *capable* of yielding knowledge gets confusedly mixed with the false thesis that information yields knowledge *in all cases*.

We maintain that, contrary to VT_N, a signal *s* being *F* can carry natural information about an object *o* being *G* even when *o* is *not G*. This is a consequence of PRT_N, which we endorse: if a signal *s* being *F* carries natural information about *o* being *G*, the probability of *o* being *G* must be higher given the signal than without the signal, but it need *not* be the case that *o* is *G*.

The core idea of a probabilistic theory of information is that signals carry natural information about anything they reliably correlate with. Reliable correlations are the sorts of correlations information users can count on to hold in some range of future and counterfactual circumstances. For instance, spots reliably correlate with measles, smoke reliably correlates with fire, and ringing doorbells reliably correlate with people at the door. It is by virtue of these correlations that one can dependably infer measles from spots, fire from smoke, and visitors from ringing doorbells.

Yet correlations are rarely perfect. Spots are occasionally produced by mumps, smoke is occasionally produced by smoke machines, and doorbell rings are occasionally produced by naughty kids who immediately run away. It follows that these paradigmatic signals do not guarantee the occurrence of, respectively, *measles, fire,* and *visitors*. They simply make such occurrences significantly more likely.

Consider the signal that the doorbell is ringing. In order for the ringing doorbell to carry the natural information that a visitor is at the door, a

---

[9] For a broader critique of counterfactual theories of information, see Scarantino 2008 and Demir 2008. For a response, see Cohen and Meskin 2008.

visitor must be at the door. This seems right. But it does not follow that a ringing doorbell carries *no* natural information about there being a visitor at the door in the absence of a visitor at the door. If the ringing doorbell is a standard one in standard conditions, it carries the natural information that a visitor is *probably* at the door. It is on account of this information, the only natural information a standard doorbell manages to carry about visitors, that we reach for the door and open it.[10]

Unlike all-or-nothing information, probabilistic natural information comes in degrees. For instance, if *o* being *G* is much more probable given the signal constituted by *s* being *F* than it would have been without the signal, *s* being *F* carries *a lot* of probabilistic natural information about *o* being *G*. If, instead, the probability of *o* being *G* increases only marginally given the signal, *s* being *F* carries a *limited amount* of probabilistic natural information about *o* being *G*.[11]

If someone wants to reduce knowledge that *o* is *G* to the information-caused belief that *o* is *G*, the limiting case of all-or-nothing information becomes central. But an account of information need not be built solely to provide a naturalistic foundation to knowledge. It may also be built to provide a naturalistic foundation for the explanation of cognition and behavior—the kind of explanation sought by current cognitive science.

What explains cognition and behavior is, more often than not, probabilistic information. Organisms survive and reproduce by tuning themselves to reliable but imperfect correlations between internal variables and environmental stimuli, and between environmental stimuli and threats and opportunities. In comes the sound of a predator, out comes running. In comes the redness of a ripe apple, out comes approaching.

The sorts of correlations on which behaviors are based commonly include some uncorrelated token events along the way—the more uncorrelated token events there are, the weaker the correlation. Organisms do make mistakes, after all. Some mistakes are due precisely to the

---

[10] Dretske (1981) suggests that not everything that could in principle make a signal equivocal *counts* as making the signal equivocal. Many background conditions (e.g., the integrity of the wires in a doorbell) can simply be assumed to be stable, and will qualify as "channel conditions." Even granting for the sake of argument Dretske's notion of "channel conditions," there appear to be innumerable nonchannel conditions that make most signals of practical interest equivocal (e.g., naughty kids ringing doorbells).

[11] Even though Dretske's (1981) theory of information focuses exclusively on all-or-nothing information, Dretske was not oblivious to the fact that natural information can come in degrees. He wrote that "[i]nformation about *s* comes in degrees. But the information that *s* is *F* does not come in degrees. It is an all or nothing affair" (Dretske 1981, 108). Thus, although Dretske's emphasis on all-or-nothing information may appear to commit him to the Veridicality Thesis for Natural Information (VT_N), he actually endorsed only the weaker thesis that if a signal *s* being in state *F* carries *the* information *that o* is *G*, then *o* is *G*. We too endorse this weaker thesis. In summary, we think that *o* must be *G* in order for a signal to carry the natural information *that o* is *G*, but that *o* need not be *G* in order for a signal to carry natural information *about o* being *G*.

reception of probabilistic information about events that fail to obtain. For instance, sometimes nonpredators are mistaken for predators because they sound like predators; or predators are mistaken for nonpredators because they look like nonpredators.

The key point is that an event's failure to obtain is compatible with the reception of natural information about its obtaining, just like the claim that the probability that o is G is high is compatible with the claim that o is not G. No valid inference rule takes us from claims about the transmission of probabilistic information to claims about how things turn out to be.

Thus, contrary to VT_N but consistently with PRT_N, a signal s being in state F can carry natural information about an object o being G even though o is not G. It follows that an agent A can have natural information about o being G whether or not o is G, provided a reliable correlation exists between the signal type received by A and the event type instantiated by o being G.

4. Nonnatural Information: The Case for the Veridicality Thesis

Talk of false information applies primarily to what we have called nonnatural information. In Grice's (1957) example, three rings on the bell of the bus carry nonnatural information about the bus being full. Suppose now that a distracted driver rings the bell of the bus three times when the bus is almost empty. In such a case, we would ordinarily speak of the three rings giving false (nonnatural) information about the fullness of the bus.

Similarly, if someone during a card game tells you that your opponent has an ace when he does not, we would ordinarily say that you have received false (nonnatural) information. We would also speak of false (nonnatural) information when a map places a hidden treasure in the wrong location. Advocates of the Veridicality Thesis recognize that people often speak of false information. But they offer a deflationary interpretation of these habits of language, concluding that, on reflection, so-called false information is not *really* information. As a consequence, they defend the Veridicality Thesis for Nonnatural Information (VT_{NN}):

> (VT_{NN}) If a signal s being F carries nonnatural information that p, then p.[12]

A corollary of VT_{NN} is that if an agent A has nonnatural information that p when she receives a signal s in state F, then p. We reject VT_{NN}, and will

---

[12] In the case of nonnatural information, the distinction between information that o is G and information that o is probably G makes no difference for our purposes, so we will work with an unspecified proposition p from here on. Whereas we have supported the thesis that a signal can carry *natural information that o is G* only if o is G (see footnote 11), we will now reject this thesis for *nonnatural information*.

argue instead that an agent A may have nonnatural information according to which p (e.g., the bus is full, your opponent has an ace, the treasure is in such and such a place) even though p is false.

The most articulate defense of VT_{NN} has been offered by Floridi (2005, 2007), who gives two main arguments in support of the thesis that false "nonnatural information" is not really information: the *argument from splitting* and the *argument from semantic loss*. We consider them in turn.

Floridi expands on an analogy originally offered by Dretske, who wrote that "*false* information and *mis*-information are not kinds of information—any more than decoy ducks and rubber ducks are kinds of ducks" (Dretske 1981, 45).[13] When we say that x is a false duck, or a false policeman, in effect we are saying that x is *not* a duck, or x is *not* a policeman. By analogy, when we say that x is false information, we might be saying that x is *not* information.

Of course, another interpretation is available. False information may be analogous to false propositions, impressions, and testimonies. As Floridi acknowledges, when we say of x that it is a false proposition, we do not mean that x is *not* a proposition. Similarly, when we say that something is a false impression or a false testimony, we do not mean that they are, respectively, *not* an impression and *not* a testimony.

So why should we conclude that "false information" works like "false duck" and "false policeman," rather than like "false proposition" and "false testimony"? Floridi suggests that "false" is used *attributively* in "false duck" and *predicatively* in "false proposition" (cf. Geach 1956), and that this explains why a false duck is not a duck whereas a false proposition is still a proposition.

Floridi (2005, 364–65) distinguishes between attributive and predicative uses of adjectives through a *splitting test*. When an adjective is used predicatively, as "male" in "male policeman," the compound can be split: a male policeman is both a male *tout court* and a policeman. When an adjective is used attributively, as "good" in "good policeman," the compound cannot be split: a good policeman is not both good *tout court* and a policeman, as bad people who are good policemen show.

Now consider a false proposition p—for example, that the earth has two moons. Floridi argues that it passes the splitting test: p is false, and p is a proposition. This being the case, "false" is used predicatively in "false proposition." Floridi suggests that, on the contrary, "false" is used attributively in "false information," because "false information" fails the splitting test: "[I]t is not the case ... that p constitutes information

---

[13] In this passage, Dretske is only referring to information in the "nuclear sense," which by definition entails truth. He adds that he is "quite prepared to admit that there are uses of the term 'information' that do not correspond to what I have called its nuclear sense" (1981, 45). Our work is in part an exploration of these other uses.

about the number of natural satellites orbiting around the earth *and* is also a falsehood" (2005, 365).

This argument from splitting is supposed to support the thesis that false information is not really information. But it requires the brute intuition that *that the earth has two moons* is not information. The content of this intuition is nothing but an instance of the general thesis to be established. Thus, the argument is question-begging. No independent reason to reject instances of false information as information is given. Whether false information passes the splitting test depends on whether we accept that a false *p* can constitute nonnatural information. We do! And so, we submit, do most other people, including most cognitive scientists and computer scientists. As far as we are concerned, *that the earth has two moons* is both nonnatural information and false.

Floridi (2007, 39–40) offers a second argument against false information. He argues that information may be lost "by semantic means." To begin with, consider two uncontroversial cases of information loss by *non*semantic means. Suppose a chemistry manuscript were burned: the information contained in the manuscript would be lost "by physical means." Or suppose that all the letters contained in the manuscript were randomly recombined: the information contained in the book would be lost "by syntactic means."

Suppose now that all true propositions in a chemistry manuscript were transformed into their negations. For instance, the book now contains sentences like "Gold's atomic number is not 79," "Water is not H_2O," and so on. Floridi maintains that the information originally contained in the book would be lost "by semantic means." But if false information— say, the information that water is not H_2O—also counted as information, Floridi thinks we would have to conclude that no information is lost.

The conclusion does not follow, however, unless we accept an unduly restrictive notion of information loss. To see why, we need to disambiguate the expression "loss of information," which can mean at least three different things. Floridi construes "loss of information" in terms of the depletion of the total amount of information contained in an information repository, such as a manuscript. Call this the *quantitative* reading of "loss of information." On this construal, if all the manuscript's true propositions are transformed into false propositions, and VT_{NN} is rejected, there is no loss of information, because the total amount of information does not change. (Following Floridi, we are assuming that the amount of true information carried by a true proposition is equal to the amount of false information carried by its negation.) In other words, if we hold that false propositions carry information and that a certain number of propositions carrying true information are replaced by an equal number of propositions carrying false information, we must conclude that the total amount of information in the repository stays the same. But, *pace* Floridi, there is nothing objectionable in this conclusion.

When we worry about information loss, we are not primarily—if at all—concerned with the *quantity* of information contained in a repository. Rather, we are generally concerned with whether an information repository carries the *same* information it originally carried. If it does not, we may also be concerned with whether the *new* information that has been gained has the same *epistemic value* as the original information. The focus here is not on the total *amount* of information in a repository but on the specific information carried by the vehicles contained in the repository: any change of information content can produce information loss, whether or not the total quantity of information remains the same.

Under a *qualitative* reading, "an information repository loses information" means that after a specific operation on it, the information-carrying vehicles in the repository no longer carry the *same* information they used to carry. Presumably, a proposition and its negation are different pieces of information. Thus, by negating a proposition, we destroy the information it carries. If we reject VT_{NN}, we can say that after negating a proposition, we now have some new information instead of the original information. But we can also say, without inconsistency, that we have lost the original information—the information carried by the original proposition.

Notice that a quantitative reading of "loss of information" fails to account for this important notion of information loss. For example, suppose that every true proposition in a chemistry manuscript were to be replaced by a *true* proposition taken from a biology manuscript, and let's assume that the amount of information carried by the pertinent chemical and biological truths is the same. In this case, there would be major information loss in the qualitative sense—the sense that matters to us in most practical contexts—even though the total quantity of information would remain just the same.

Finally, under an *epistemic-value* reading, "an information repository loses information" means that after a specific operation on it, the information-carrying vehicles in the repository carry information with a lower epistemic value. Obviously, in this sense there can be information loss even if we reject VT_{NN}. Since false information is epistemically inferior to true information, negating true propositions is a way to incur information loss by semantic means.

Summing up, rejecting VT_{NN} is compatible with accounting for information loss "by semantic means" in the two senses—the qualitative and epistemic-value senses—that matter most for epistemic purposes. Moreover, our distinctions allow us to neatly distinguish between physical and syntactic information loss on the one hand and semantic information loss on the other. In the first two cases, information is destroyed but not replaced with any new nonnatural information. There is information loss in the quantitative sense, in the qualitative sense, and in the epistemic-value sense. In the third case, information is destroyed and replaced with new (false, and thus epistemically inferior) nonnatural information. There is information loss in the qualitative sense and in the epistemic-value sense, though not in the quantitative sense.

5. Nonnatural Information Without Truth

Our main reason to maintain that nonnatural *false* information is information too mirrors our reason to posit nonnatural information in the first place: it allows us to capture important uses of the term "information." It is only by tracking such disparate uses that we can make sense of the central role information plays in the descriptive and explanatory activities of cognitive scientists and computer scientists, which partially overlap with the descriptive and explanatory activities of ordinary folk.

As we did for natural information, we are trying to understand the kinds of information that are actually invoked in the explanation of cognition and behavior. In the case of *natural information*, the need to capture the explanatory power of the notion of information requires positing the notion of *probabilistic information*. In the case of *nonnatural information*, the need to capture the explanatory power of the notion of information requires positing *false* information and *non-truth-evaluable* information.

5.1. False Information

Cognitive scientists routinely say that cognition involves the processing of information. Sometimes they mean that cognition involves the processing of natural information. At other times, they mean that cognition involves the processing of nonnatural information, without any commitment as to whether the information is true or false. For instance, a recent study on how aging affects belief found that "[o]lder adults were more likely than young adults to believe false information and their dispositional ratings were reliably biased by the valence of false information" (Chen 2002, 217). Statements to the effect that nonnatural information can be false are widespread in many sciences of mind.

If we accept VT_{NN}, we should endorse an *error theory* to make sense of this way of talking. According to such a theory, cognitive scientists are mistaken about the nature of cognition. Since false nonnatural information is not *really* information (as per VT_{NN}), the view that cognition involves information processing is at best incomplete. It should be supplemented by the view that, unbeknownst to most cognitive scientists, cognition also involves the processing of *misinformation*, where misinformation is understood as different in kind from information. This error theory of cognitive science is uncharitable, awkward, and unnecessary.

The notion of nonnatural information used in cognitive science is best interpreted as the notion of *representation*, where a representation is by definition something that can get things wrong. The sentence "Water is not H_2O" gets things wrong with respect to the chemical composition of water, but it does not fail to represent that water is not H_2O. By the same token, the sentence "Water is not H_2O" contains false nonnatural information to the effect that water is not H_2O.

One of the central challenges for a theory of cognition and behavior is to explain how a system can acquire the ability to represent, or nonnaturally mean, or carry nonnatural information, that p whether or not p is true. There is a large literature devoted to explaining how this works. Our present objective is not to explain how representation works or whether representation can be naturalized using the notion of natural information. Our objective is simply to capture existing usage and distinguish the commitments pertaining to different notions of information. Nonnatural information as used in cognitive science has the capacity to get things wrong, so it can be false.

Computer scientists are another group of scientists who refer to nonnatural information independently of whether it is true. Computer scientists routinely label as information processing all the cases in which computers process semantically evaluable structures, whether they are true or false. If we accept VT_{NN}, we should conclude that computer scientists are also mistaken: computers are more than information processors—they are misinformation processors as well. But again, this *error theory* of computer science is uncharitable, awkward, and unnecessary. It is clear that computer scientists include both true and false information under the rubric of (nonnatural) information. This is yet another reason to conclude that false information is information too.

Thus, contrary to VT_{NN}, a signal s being in state F can carry nonnatural information that p even though p is false. It follows that an agent A can have nonnatural information that p whether or not p.[14]

5.2. Non-Truth-Evaluable Information

We have so far focused on declarative nonnatural information, the sort of information paradigmatically associated with declarative sentences. Declarative nonnatural information is truth-evaluable—it may be either true

---

[14] Fetzer (2004) has also argued that nonnatural information can be false, but his arguments are problematic (see Sequoiah-Grayson 2007). Fetzer (2004, 226) objects that certain sources of information, such as blood spots, PET scans, and tree rings, may "lack a propositional or a sentential structure." But these classes of signals can easily be given a sentential structure if we understand them as events constituted by particulars having properties (e.g., a tree having thirty-three rings constitutes an event). Also, Fetzer (2004, 225) presents sentences such as "There is life elsewhere in the universe" as counterexamples to VT_{NN}, on the grounds that the truth-value of such sentences is unknown and may never be known. But our ignorance is not evidence against VT_{NN}. Given VT_{NN}, "There is life elsewhere in the universe" is information just in case there is life elsewhere in the universe. Whether or not we know it is irrelevant. On our view, "There is life elsewhere in the universe" would be information even if we knew that there is no life elsewhere in the universe.

or false. We have argued that false declarative nonnatural information is information too. There are also forms of nonnatural information to which the very categories of truth and falsity do not apply.[15]

Tracking the uses of "information" in cognitive science and computer science demands considering forms of nonnatural information that are not truth-evaluable. Consider giving people information on what to do. If you tell us to buckle up, you give us information. You inform us on what to do. But what you say is not true—not because it's false, but because it's not truth-evaluable.

Commands are just one example of nondeclarative speech acts. Other nondeclarative speech acts include advising, admonishing, forbidding, instructing, permitting, suggesting, urging, warning, agreeing, guaranteeing, inviting, offering, promising, swearing, apologizing, condoling, congratulating, greeting, and many others. The point of such speech acts is to transfer information, but not for the purpose of describing something. Rather, the purpose of transferring nonnatural nondeclarative information may be to express a certain emotional state ("I am sorry"), to commit to doing something ("I promise"), to get someone to do something ("Buckle up"), to produce a certain state of affairs ("I pronounce you husband and wife"), and so on. Requiring nonnatural information to be true would prevent us from understanding nondeclarative speech acts as having anything to do with information transfer.

A similar point can be made with respect to the use of "information" in computer science. A core insight of modern computer science, which made possible the design of universal digital computers,[16] is the equivalence of *data* and *instructions*. Computers operate on data on the basis of instructions. Both data and instructions are physically instantiated as strings of digits, which computers can manipulate. In other words, the data and instructions that computers manipulate are intrinsically the same kind of thing—they differ only in the role they play at any given time within the computer. The very same string of digits may play either the role of instruction or that of data at different times.

Computer scientists usually assign computer data a semantic interpretation—data can have semantic content. Thus data can carry informa-

---

[15] We emphasize that we are *not* accusing supporters of VT_{NN} of ignoring that nondeclarative sentences are not truth-evaluable. Floridi (2005) explicitly states that his theory of "strongly semantic information" applies only to declarative information. Moreover, his notion of "instructional information" (Floridi 2008) seems to overlap to a large extent to what we call non-truth-evaluable information. Nonetheless, non-truth-evaluable information provides further uses of "information" that do not presuppose truth. Given that both declarative and nondeclarative information are covered by the theses that cognition involves information processing and that computing systems process information, this is further evidence that VT_{NN} does not apply to "information" as used in cognitive science and computer science.

[16] Roughly speaking, a computer is universal just in case it can compute any function computable by an algorithm (until it runs out of memory).

tion. Computer instructions are usually interpreted semantically too—they are assigned as semantic content the operations the computer performs in response to them. But instructions are neither true nor false, because they are not truth-evaluable at all. Yet, there is no intrinsic difference between instructions and data. In addition, both have semantic content. If data can carry information, so can instructions.

The existence of non-truth-evaluable information provides another plausibility argument for false information. Given that non-truth-evaluable information carried by computer instructions and nondeclarative speech acts is information, and that the information carried by true data and true declarative information is information, it's hard to see why the information carried by false data and false declarative sentences should not be information too.

6. An Objection

At this point, a supporter of the Veridicality Thesis (VT)—which we understand from here on as the conjunction of VT_N (section 3) and VT_{NN} (section 4)—may be tempted to object as follows (cf. Sequoiah-Grayson 2007).[17] Information is a mongrel concept, as are many concepts taken from ordinary language. Subjecting a theory of information to critique because it fails to capture *some* usages of "information" presupposes that a good theory of information ought to capture *all* usages. But no theory of information can achieve that objective, precisely because the concept of information is a mongrel.

According to this objection, the notions of probabilistic natural information on the one hand and false and non-truth-evaluable nonnatural information on the other show only that theories of information that abide by VT are not all-encompassing. But this conclusion is trivial, so the account we have offered in this chapter is not especially interesting.

The objection would be well taken if our critique were simply that there exist ordinary usages of "information" that violate VT. This alone would be fairly inconsequential. But our point is different: any theory of information interested in making sense of the central descriptive and explanatory role information plays in both the sciences of mind and computer science must give up both VT_N and VT_{NN}. This is quite significant.

We are emphatically *not* proposing that a unique theory of information is forthcoming. This is not why we find VT problematic. In fact, we have taken pains to distinguish two importantly different usages of "information," and two importantly different versions of VT, namely, VT_N and VT_{NN}.

[17] Sequoiah-Grayson's (2007) objection emerges in response to some earlier critiques of the thesis that information entails truth (e.g., Devlin 1991, Fetzer 2004). See also Floridi (2005) on Fox's (1983) and Colburn's (2000) critiques.

To make sense of natural and nonnatural information, we need different theories of information. We need a theory of the conditions under which signals carry natural information. We also need a theory of the conditions under which signals carry nonnatural information. And, ideally, we'd like a theory about how organisms manage to manufacture signals carrying nonnatural information out of (among other things) signals carrying natural information.

In other words, our rejection of VT is consistent with the pluralism we advocated in section 2. What we have argued is that "information without truth," in the various senses we have distinguished, can help us understand the central role played by both natural and nonnatural information in the description and explanation of cognition and behavior.

What we find at odds with pluralism is precisely the idea that VT should be one of the "basic principles and requirements for *any* theory of semantic information [and] that false information fails to qualify as information at all" (Floridi 2007, 40, emphasis added). On the contrary, we have argued that there are legitimate notions of semantic, factual, epistemically oriented nonnatural information that qualify false information as genuine—though epistemically inferior—information. And we have argued that there are legitimate notions of semantic, factual, epistemically oriented natural information that qualify probabilistic information as genuine information, even though it is not the sort of natural information of which knowledge is constituted.

Finally, we are not suggesting that a theory of information should be rejected just *because* it incorporates some version of the Veridicality Thesis. As long as there are legitimate theoretical objectives fulfilled by the theory—from the naturalization of knowledge to the solution of certain puzzles[18]—endorsing a truth requirement on information need not be a liability. What we do reject is that VT belongs in an illuminating account of the central role played by natural and nonnatural information in the explanation of cognition and behavior.

7. Conclusion

We have discussed whether the primary notions of information used in cognitive science and computer science, namely, natural information and nonnatural information, commit us to the thesis that information entails truth. We have argued that the thesis does not hold in its general form for either natural or nonnatural information.

[18] For instance, Floridi's (2005) theory of information provides a solution to the so-called Bar-Hillel-Carnap Paradox. The paradox emerges if we follow Carnap and Bar-Hillel (1952) in holding that the information content of a sentence corresponds to the set of possible worlds excluded by its truth. On this assumption, contradictions have maximal information content, because they exclude all possible worlds.

Natural information can be carried by reliable but imperfect correlations. Thus, a token signal may deliver probabilistic natural information to the effect that, say, someone is at the door even if no one actually is. The transmission of natural information, we have concluded, entails nothing more than the truth of a probabilistic claim.

The notion of (declarative) nonnatural information is best understood as the notion of representation, which by definition can be false. Consequently, the sentence "The earth has two moons" carries nonnatural information just like the sentence "The earth has one moon." At the same time, we are not saying that true and false nonnatural information are on an equal epistemic footing: true information is epistemically more valuable than false information.

Finally, we have argued that nonnatural information need not be truth-evaluable at all—the information contained in nondeclarative speech acts and computer instructions is neither true nor false, because it is not truth-evaluable.

Our main rationale for rejecting the Veridicality Thesis in the two versions we have considered is that the thesis stands in the way of our understanding of the role played by information in the descriptive and explanatory efforts of cognitive scientists and computer scientists. We have emphasized that defining a specialized notion of information that entails some version of the Veridicality Thesis may still be useful for some specific purposes. Our point is that such a specialized notion should not be understood as germane to the main notions of information used in cognitive science and computer science.

Acknowledgments

We are grateful to Fred Adams, Patrick Allo, and Hilmi Demir for comments on a previous draft.

References

Carnap, Rudolf, and Yehoshua Bar-Hillel. 1952. "An Outline of a Theory of Semantic Information." Massachusetts Institute of Technology Research Laboratory of Electronics, Technical Report No. 247. Cambridge, Mass.: Massachusetts Institute of Technology.

Chen, Yiwei. 2002. "Unwanted Beliefs: Age Differences in Beliefs of False Information." *Aging, Neuropsychology, and Cognition* 9:217–28.

Cohen, Jonathan, and Aaron Meskin. 2006. "An Objective Counterfactual Theory of Information." *Australasian Journal of Philosophy* 84:333–52.

———. 2008. "Counterfactuals, Probabilities, and Information: Response to Critics." *Australasian Journal of Philosophy* 86:635–42.

Colburn, Timothy R. 2000. "Information, Thought, and Knowledge." *Proceedings of the World Multiconference on Systemics, Cybernetics and Informatics* X:467–71 (Orlando, Fla., July 23–26, 2000).

Demir, Hilmi. 2008. "Counterfactuals vs. Conditional Probabilities: A Critical Examination of the Counterfactual Theory of Information." *Australasian Journal of Philosophy* 86:45–60.

Devlin, Keith. 1991. *Logic and Information*. Cambridge: Cambridge University Press.

Dretske, Fred. 1981. *Knowledge and the Flow of Information*. Oxford: Blackwell.

———. 1988. *Explaining Behaviour*. Cambridge, Mass.: Bradford/MIT Press.

Fetzer, James H. 2004. "Information: Does It Have to Be True?" *Minds and Machines* 14:223–229.

Floridi, Luciano. 2005. "Is Semantic Information Meaningful Data?" *Philosophy and Phenomenological Research* 70:351–70.

———. 2007. "In Defence of the Veridical Nature of Semantic Information." *European Journal of Analytic Philosophy* 3:31–42.

———. 2008. "Semantic Conceptions of Information." In *The Stanford Encyclopedia of Philosophy*, edited by Edward N. Zalta. URL http://plato.stanford.edu/archives/win2008/entries/information-semantic/

Fodor, Jerry A. 1990. *A Theory of Content and Other Essays*. Cambridge, Mass.: MIT Press.

Fox, Christopher J. 1983. *Information and Misinformation: An Investigation of the Notions of Information, Misinformation, Informing, and Misinforming*. Westport, Conn.: Greenwood Press.

Geach, Peter T. 1956. "Good and Evil." *Analysis* 17:33–42.

Grice, Paul 1957. "Meaning." *Philosophical Review* 66:377–88.

Kim, Jaegwon. 1976. "Events as Property Exemplifications." In *Action Theory*, edited by Miles Brand and Douglas Walton, 159–77. Dordrecht: Reidel.

Lewis, David. 1973. *Counterfactuals*. Oxford: Basil Blackwell.

Millikan, Ruth G. 2004. *Varieties of Meaning: The 2002 Jean Nicod Lectures*. Cambridge, Mass.: MIT Press.

Piccinini, Gualtiero, and Andrea Scarantino. Forthcoming. "Computation vs. Information Processing: Why Their Difference Matters to Cognitive Science." *Studies in the History and Philosophy of Science*.

Scarantino, Andrea. 2008. "Shell Games, Information, and Counterfactuals." *Australasian Journal of Philosophy* 86:629–34.

Scarantino, Andrea. Unpublished. "A Theory of Probabilistic Information."

Sequoiah-Grayson, Sebastian. 2007. "The Metaphilosophy of Information." *Minds and Machines* 17:331–44.

Shannon, Claude E. 1948. "A Mathematical Theory of Communication." *Bell System Technical Journal* 27:379–423 and 623–56.

Stampe, Dennis. 1975. "Show and Tell." In *Forms of Representation: Proceedings of the 1972 Philosophy Colloquium of the University of Western Ontario*, edited by Bruce Freed, Ausonio Marras, and Patrick Maynard, 221–45. Amsterdam: North-Holland.

——. 1977. "Toward a Causal Theory of Linguistic Representation." In *Midwest Studies in Philosophy*, volume 2, edited by Peter French, Theodore Uehling, and Howard Wettstein, 81–102. Minneapolis: University of Minnesota Press.

6

INFORMATION AND KNOWLEDGE À LA FLORIDI

FRED ADAMS

Information and Meaning

In "Open Problems in the Philosophy of Information," Floridi (2004a, 560) lists five "extensional" approaches to the notion of information, ranging from Shannon and Weaver's and Dretske's notions of semantic information cast in terms of probability space and the inverse relation between amounts of information and probability of a proposition *p*, to tracking the possible transitions in state space or tracking possible valid inferences relative to a person's epistemic states. While there are differences in the approaches, there are similarities as well. In what follows, I'm going to take the path that Floridi himself charts—that of defining information in terms of a data space where information is semantic, well formed, meaningful, and truthful data. Floridi's approach is perfectly consistent with the approaches of Shannon and Weaver and of Dretske as well, and where there may be differences, those differences do not matter for the topics I discuss below.[1]

In my view, information is a naturally occurring commodity. It exists independently of minds and is created and transmitted by the nomic regularities of occurrences of events (and their properties). Whether there is information independently of the world of physical events, I do not know. Floridi (2004a, 573) poses the question "Can information be naturalized?" My answer is yes. He also asks whether *environmental information* requires higher-level cognitive representations. My answer is no. Purely physical events can create information by their occurrence and be transmitted via their nomic connection with other events. But can higher-level cognitive representations create, emit, and encode information? Yes, indeed. If they could not, this collection of chapters would not exist. Is it the same information (quantity and content) that exists in environmental information and in cognitive information? It can be. When a metal bar expands, its expansion carries information about an increase in the local ambient temperature. When a cognitive agent learns of that

---

[1] Of course, Shannon does not himself consider using information as a basis for developing semantics. Dretske (1981) does, but his use of information has modal and counterfactual properties focused upon the origin of concepts and beliefs that derive from information. Floridi is developing a semantic notion of information. See my 2004 for more.

fact (the bar is expanding) and knows of the connection between expansion and temperature, the agent learns the fact that the ambient temperature is rising (if the agent did not already know this). Is it the *same information* in the environmental events and in the mind of the cognitive agent? Yes. Indeed, it had better be the same information, if knowledge does what we think it does—connect us to the world. Knowledge is *of the world*. Therefore, knowledge is *of the truth*. And if *information* is the bridge between the cognitive agent and the world, and the truth, then it had better be the exact same commodity and content that the cognitive agent *learns* when the agent learns information *about the world*. In my view, and I think in Floridi's, this requires information to be a naturally occurring commodity that can be transmitted and exploited by cognitive agents. And one key is that the mind of a cognitive agent must be able to exploit the information it acquires and transforms into meaning and understanding—without the need of another mind to impose meaning or interpretation on that content. This is what Searle (1983) has called *intrinsic* content and others (Adams and Aizawa 2005, 2008) have called *nonderived* content.

Now, can noncognitive organisms exploit information? Yes. Trees exploit information about the amounts of sunlight, when they stop growing and drop their leaves. Do they cognize this information? No. They don't need do. Their exploitation is of a purely noncognitive variety. Shorter days promote reduction in photosynthesis. Reduction in photosynthesis promotes reduced flow of sap and reduced growth. Reduced flow of sap and reduced growth (dormancy) promotes withering of and dropping of leaves—all without need or benefit of consciousness or cognition.

Indeed, it turns out that even more interesting uses of information are possible. The acacia tree has the capacity to increase its output of tannin. When kudu antelope eat the bark of the tree, this prompts the tree to increase its output of tannin, even in release from its leaves that is carried downwind in the air, as well as in the production of bark. Since kudu do not like high-tannin fodder, this discourages them from eating the acacia or foraging downwind. So, once again, information is exploited (about damage to the bark) to produce an effect that is beneficial (increased tannin production, reduced destruction by kudu). As before, this is noncognitive exploitation of information. Now if you like, one could also say that this is all explainable in purely physical terms of environmental causes and reaction of the tree. And that is so. Still, it counts as information because tracking these events and their interrelations could bring an intelligent agent to know (it is time to drop leaves or it is time to increase tannin). The trees are not intelligent agents and do not know. But that does not mean that the events in the causal chain do not contain information or that the events which benefit the trees do not exploit the informational connections (regularities) in nature that exist.

What about lower-level living systems—amoebas, paramecia, bacteria? Once again, these systems exploit information about their environments in purely noncognitive ways. This does not mean that there is no informational *intentionality* or *aboutness* in the information being exploited (Fitch 2008), but it is not a level of intentionality sufficient for cognition—for example, *misrepresentation*, higher orders of *intensionality* (Dretske 1981). Still, to explain many of the kinds of information that lower biological systems exploit, one does need to appeal to a selectional history to explain the exploitation of information in the environmental niche of the type of organism (Dretske 1995, 156).

It is a further interesting question whether there can be purely sensory systems with phenomenal qualitative states, but no cognitive states. Suppose a creature could detect and phenomenally feel a warm or cool surface or a flat or rough surface. Suppose further that the creature had no concepts of warm or cool, flat or rough. Such creatures would still process information about the environment or about their own bodies or the interaction of their bodies with their environments, but would be void of concepts and cognition. These would be systems that live in a world of *pure feel*. They would still exploit information in the sense that events occurring in their bodies carried information about the environment (warm here, cool there, smooth surface here, rough surface there, damage to body here), though these represented events would not be *represented as* warm, cool, smooth, or rough surfaces. They could not be, or the system would not be a purely sensory system. To have concepts and be able to conceptualize these states would constitute being a fully cognitive system. I say it is an interesting question because many writers who have discussed the matter of the informational (representational) view of *qualia* (Dretske 1995, 19; Tye 1995, 144; Evans 1982, 158; Carruthers 2005) have all suggested that only when a system with sensory capacity is harnessed to a cognitive system (with concepts) do truly phenomenal states arise. I suspect that purely sensory systems are possible or even do exist, though I can't say whether they would qualify as having *minds*. Somewhat surprisingly, those like Dretske, Evans, and Tye who have claimed that such systems would not have phenomenally qualitatitve states have not given arguments why not.

Why don't amoebas and paramecia have qualitative experiences? Fitch (2008) would say that although they have nano-intentionality, they don't have internal mental models. Evans (1982), Dretske (1995), and Tye (1995) would say they don't have sensory systems that feed into conceptual systems. Even though all would agree that information is processed (at the first level of intentionality where even if a signal carries the information that *a* is F and as a matter of fact all Fs are Gs by coincidence, the signal will not thereby carry the information that *a* is G. There is a perfectly good sense in which these organisms don't need qualitative experiences. The differential responses that they make to environmental changes do amount to a low-level processing of information (Fitch calls it *nano-intentionality*), but the processes involved can all be explained at the level of chemistry (or photochemistry) alone (plus a bit of the selectional history of the organism in its local environment). There are no *dedicated processors*—no biologically selected structures that are recruited for the purpose of tracking information about the environment or the internal states of the organism as they respond to environmental changes. Thus, there are no more or less permanent structures that have the biological function of indicating to (tracking for) the organism changes of environment and self. Thus there are no internal structures that have the function of both informing a cognitive system and driving a motor system that serves the needs and desires of the organism. Having such internal, dedicated information processing structures requires explanation that rises above the level of local chemical reactions. So it may well be that an organism's use of the information generates a qualitative sensory experience, and without such, there are no qualitative states.[2]

My interest is in how one moves from an account of information to an account of cognition and meaning that can support knowledge and purposive behavior. To support an account of knowledge, there must be a special connection between information and truth. This connection is a basic feature of Floridi's (2005) notion of information, viz. information that *p* cannot be false. To support an account of cognition and purposive behavior, there must be an explanation of how information can contribute to representations in a system capable of meaning, a type of meaning that can be false or represent future states to be brought about. If information is meaningful data and cannot itself be false or about the future, but thoughts, beliefs, intentions and the like can be false (purposive systems can make mistakes), then there must be transformations that information must undergo for it to provide a resource in the development of states which have meaning and which can be falsely tokened. A purposive creature can enter a state that means *that p is the case*, when *p* is not the case. Information cannot do this, if Floridi is right (and I think he is). So what has to happen for something that cannot be false (information) to play a constitutive role in the origin of something that can be false, such as a belief, a thought, or a statement?

Now why is the explanation of this transition an important matter to solve? If you were going to build a mind, what would you have to put into the system? If the mind you are building is not going to be a sensory system only, a system that *feels* but does not *think*, then you will need to put in a *language of thought*. The system will need a set of symbols with which it can *track* the world and its movements through it. At times it will need to make its behavior conform to the world to satisfy its needs for survival and prosperity. At other times it will need to make the world conform to its desires and needs. In order to do this, it will need to be able to think. Its thought will require symbols that are about the world

---

[2] See Adams and Beighley forthcoming and Adams unpublished for more discussion.

and that can concatenate into propositions about it and its place in the world. And where will these symbols get their meaning? Indeed, what will it be for these symbols to *have* a meaning or be *about* the world or about the system itself?

Much of Floridi's work has been to construct and defend a concept of *information* that meets the needs of a theory of *semantics* that can account for how the symbols in a language of thought acquire their *contents* or *meanings*. Of course, Floridi is not alone in this. He has shared the goal with others who have come before him with the same goal and same general ideas about what is needed (Dretske 1988, Fodor 1990, Millikan 1989, and others). Perhaps what is distinct about Floridi's contribution to the project of naturalizing the mind and naturalizing semantics is his focus on and determination of the needs of the concept of *information* to suit this purpose. What properties must *information* bear to serve the purpose of explaining how symbols in the language of thought acquire their meanings from their *informational roots* or sources? This is the part of Floridi's work on information that interests me the most and that I will discuss here.

It is a semantic notion of information that is Floridi's quest. As for Dretske (1981) before him, for Floridi "the sort of 'information' that interests us here is arguably the most important. ... It is 'information' as semantic content that ... allows the elaboration of an agent's propositional knowledge" (2006, 3). So there is a sense of information that *p* or about state of affairs *f* that exists in one's cognitive system (one's beliefs, or perceptions or knowledge). This sense of information is semantic—it has a truth value. Indeed, Floridi argues (2005; 2006, 10) that its truth value has to be *true* to be genuine information that *p*.

There is also information in the sense of *natural sign* or *nomic regularity*, where information can exist outside cognitive agents. Indeed, were it not so, how could there be information to be delivered to the cognitive states of cognitive agents? The information that cognitive states possess must have an origin, must come from somewhere. From where, if not from naturally occurring events in one's environment? For such events to carry information they must occur against a background of environmental regularity and consistency in which one event's occurrence is so regularly and lawfully connected with another (as a kind) that the one event's occurrence is capable of bringing one to know of the other event's occurrence. This ability to bring about knowledge is supported by the objective dependence between the two event types, the dependence of the one's occurrence upon the other's. Expanding metal carries information about rising temperature. Tracks in the snow carry information about passersby. Rings in a tree carry information about seasons of growth. But the sense of information here is that of natural signs of things signified, not of something that has a meaning in the sense of being able to be evaluated for truth or, more important, for falsity. Semantically interpreted symbols can say something false. Natural signs cannot.

The holy grail in artificial intelligence research is to build systems that not only pick up information in the sense of natural signs of the environment around them (or of their own states) but also are able to transform that information into symbols that are meaningful to the systems themselves, not just to their builders. These systems must store or manipulate not mere *uninterpreted symbols* or what Foridi calls *mere data* but symbols that are semantically informative to the agent itself. As Floridi puts it in "Information Inspiration": "[W]henever the behaviour in question is reducible to a matter of transducing, encoding, decoding, or modifying uninterpreted data according to some syntax, computers are likely to be successful. That is they are rightly described as purely syntactic machines" (2004c, 60). Computers can handle what Floridi (2004c) calls *proto-semantic* data. They can detect identities as equalities and differences, but these acts still are "far too poor to generate anything resembling semanticisation." But how can the symbols in a mind (or a computer system we are trying to turn into a mind) solve what Harnad (1990) calls the *symbol grounding problem*? How can the symbols rise to the level of meaning (with the possibility of false tokening) *and* how can they be meaningful *to the system itself*, not just to the builders of the system or the programmers?

It is my view that these two tricks need to be turned together, in one fell swoop. The trick has to be turned, as Floridi (2005) puts it, *autonomously* and *from scratch*. And the solution must be *naturalized*. We both agree on all of this. And we agree that the content a system achieves must be nonderived. But we may disagree on the sense of "nonderived." For me this means non-semantically derived. That is, no part of the meaning of a symbol or concept can have its meaning semantically depend on meaning of mine or a programmer's or an engineer's. But let's distinguish causal derivation from semantic derivation. When I teach someone what a whale is, I may show her lots of whales. I cause her to get the idea, form the concept, of a whale. This is a process of causal derivation, and I enter the process. If the child mistakes a shark for a whale, I correct her—give her negative feedback. But when the child forms a mental symbol of whales and forges the connection between that symbol and the property it represents, viz. *being a whale*, then even though I had a causal role in this, I do not have a semantic role. Her symbol means whale not because of my meaning whale by my symbol or my intending that she acquire a symbol for whale, or by a convention imposed or an interpretation imposed by me. No. Her whale-concept is formed autonomously and from scratch from the resources available in her head. I just helped causally, not semantically. There was causation between the symbol in her head and the property of being a whale, but there was not meaning in the chain prior to the meaning connection's being forged. There was *causal* but not *semantic* derivation (Aizawa and Adams 2005).

What would *semantic* derivation look like? We have cases. Words of natural language get their meaning from the contents of the concepts and intentions and conventions of the originators and speakers of the language. When a new term is introduced into the language, *titanium,* it is because the physicists and chemists could think of that metal and name it that the contents of their thoughts gave meaning to the term "titanium." The word semantically derived its content from the content of the concepts of the namers. Or consider a red stop light. In and of itself a red light means nothing about traffic. But we decide to give its shape and its color a significance (meaning to stop). Its meaning as a traffic signal derives its content from the content of our traffic planners and the minds of those drivers who adopt the convention as part of the rules of the road. In these examples, there is more than *causal* derivation. The semantic content of the word or the stop light derives its content from already existing content in the minds of the originators. The meaning of the word or the stop light *derives* from the meaning in the minds of the originators. In the case of mere causal derivation, the meaning of the symbol is causally situated by the engineers, but the content is not derived from the contents of their minds. It is un-derived in that sense. The content comes from the properties in the world to which the symbol becomes dedicated—no matter how that takes place causally.

Now this is important because Floridi seems to follow Dennett in wielding his axe against the possibility of semantic nonderivation when he sees causal derivation. In his criticism of what he calls Harnad's *hybrid* account of the symbol grounding problem, Floridi criticizes Harnad for allowing the use of back propagation techniques into his model. He says such procedures are *supervised* and so whatever grounding is *entirely extrinsic* (meaning derived or not intrinsic to the system itself). Now I don't want here to defend Harnad's particular model. Instead, I want only to point out that so far we have *causal* derivation but not necessarily *semantic* derivation. Even if Harnad used back propagation to train a neural net to be selectively sensitive to the inputs about the property of something's being a whale, the trainers merely causally help forge the connection between elements in the net and the property of inputs that carry information about whales. What the system then does about this is another matter. How the system came to have a dedicated symbol for *whale,* if that is what it has, is another matter, as well. But that it has a symbol for whale will not be a matter of convention, our interpretation, and so on. If it is a genuine symbol for whale, then to the mind of the net, it will mean *whale.* It is not merely that the trainers want it to mean that. It genuinely means that. It can be falsely tokened. It can allow the system to want a whale it doesn't have (if it has the capacity to want and think). There is no inconsistency in a symbol's meaning being both *causally derived* and *semantically un-derived.* I think Floridi is overlooking this.[3]

---

[3] Interestingly, Floridi and Taddeo do recognize a "zero semantic commitment" condition on solving the symbol grounding problem—viz. that no semantic resources be

Information and Knowledge

For some, including tracking theorists of knowledge (Adams 2005), knowledge is a real-world relation. If Colleen knows that Obama won the 2008 U.S. presidential election, then Colleen is related to the world in a special way—the knowledge way. She is connected to the truth (reality) in a special way that only knowledge (not justified belief or even justified true belief) can provide. Having secured knowledge, Colleen (via her sources of knowledge) has eliminated possible sources of error that mere justified belief and even justified true belief have not eliminated. Indeed, Floridi (2004b) has offered a proof that three factor theories alone (truth, plus belief, plus justification) cannot eliminate the possible sources of error that permit accidentally true, justified belief (that permit Gettierization).

Tracking theories offer something besides justification (if that is not itself a component of tracking) that eliminates these sources of error that could circumvent knowledge. Of course, not everyone agrees. Elsewhere I've tried to respond to the objections to tracking theories that are most prevalent in the literature (Adams 2004, 2005, Adams and Clarke 2005). Information-based accounts of knowledge (Dretske 1981, Adams 2004) are tracking theories. They exploit the very feature of information that Floridi has fought so valiantly to defend—its connection with truth. If a signal carries the information that p, then p is true. So if Colleen receives the information that Obama won, then Obama won. And that entailment is a nomic entailment. Colleen's true belief, since sustained by the information that Obama won, is not accidentally true (justified or not). Therefore, only a semantic theory of information with the properties Floridi defends (semantic, veridicality condition) fits the bill of what an information-based account of knowledge requires to avoid the problem Gettier uncovered those many years ago.

Can tracking theorists go further in acceptance of Floridi's analysis of the connection between information and knowledge? There is one extension of Floridi's account that tracking accounts may need to resist. In "The Logic of Being Informed" (2006), among the many things that Floridi defends are the following eminently plausible principles of an informational logic *a is informed that p* (a logic Floridi introduces to extend logics such as doxastic *a believes that p* or epistemic logics, *a knows that p*): *veridicality thesis* (*a* is informed that $p \rightarrow p$), *entailment property* (*a* knows that $p \rightarrow a$ believes that p), and the *distributive thesis* (if *a* is informed that $p \rightarrow q$ and is informed that p, then *a* is informed that q). Floridi follows the introduction of the distributive thesis with this: "Note that, although this is entirely

uploaded from outside the system (2005 and 2009), but (in both of these papers) seem to think this is violated by causal derivation. See Aizawa and Adams 2005 and Adams and Aizawa 2008 for much more explanation of a similar mistake that Dennett makes. And see Adams 2003 for my account of how the right story of naturalized semantics likely goes. I believe that Fodor and Dretske are on the right track about the path from information to meaning—though of course I don't agree with everything in their views.

uncontroversial, it is less trivial. Not all 'cognitive' relations are distributive. 'Knowing,' 'believing,' and 'being informed' are, as well as 'remembering' and 'recalling.' . . . However, 'seeing', and other experiential relations, for example, are not; if an agent a sees (in a non-metaphorical sense) or hears or experiences or perceives that $p \rightarrow q$, it may still be false that, if a sees (hears etc.) p, a then also sees (hears etc.) q" (2006, 8).

Floridi's choices of *exceptions* to the principle are telling. For tracking theorists of empirical knowledge, how can the knowledge resting upon perceptual states obey this principle, when the very experiential states themselves do not obey the principle? Knowledge acquires its properties from its informational base. If the experiences upon which one's perceptual knowledge rests do not obey this informational principle, then empirical knowledge itself may also fail to obey the principle.

As is well known, Nozick (1981) and Dretske (1970, 1971, 1981, 2005) reject closure for knowledge. While Nozick himself did not couch his claims that knowledge is not closed in terms of information, Dretske certainly did. Dretske maintains that one may know that p because one is informed that p. One may know that $p \rightarrow q$ because one is informed that $p \rightarrow q$, and still one may not know that q because one may still fail to be informed that q. How could this be?

As Floridi would agree, a system can send a signal that carries the full information that p only against a background of channel conditions that bar the local nomic possibility of equivocation in one's local circumstances. Now for a to receive the signal that *informs a of p*, it is *not necessary* for a to be informed that *it is information that a is receiving*. That is, it is not necessary for a to receive information about those channel conditions. In fact, it is impossible for a system to send *by the very signal that sends the information that p* simultaneously to send the signal *that this is indeed information that p*, and not an equivocal signal. To know the latter one would need other means of knowing, other signals, or other investigation of the system's channel conditions.

Indeed, Floridi explains this very well himself in explaining why his informational logic is not committed to the KK thesis (if one knows then one knows that one knows) or BB thesis (if one believes then one believes that one believes) saying "it is perfectly acceptable for a to be informed that p while being (even in principle) incapable of being informed that a is informed that p" (2006, 12). So now we can proceed to examples.

Dretske's classic example is that one may go to a zoo and see a zebra in the zebra cage, know it is a zebra by its look, know that if it is a zebra, it is not a painted mule, but not know (by its look) that it is not a painted mule. Why not? Because the look does not carry the information that it is not a painted mule. To do that, it would have to tell the knower that it was information that it was a zebra (and not a painted mule). That is, it would have to tell the knower that the channel condition of there being no pranksters running the zoo is met. But the mere look of the zebra does not, of course, tell one that. As Floridi allows, one may *see or hear* the zebra in the cage, may know that zebras are not painted mules, but not *see or hear* the animal's not being a painted mule.

Take another example. Chris uses litmus paper to tell the liquid in the beaker is an acid. Were it not an acid, the paper would not turn pink. The background condition that allows the litmus test to convey the information that the liquid is an acid is *the fact that* (if it is a fact), there are no nonacids in Chris's portion of the world that turn litmus paper pink. Of course, though this has to be true for Chris to learn by the litmus test that it is an acid in the beaker, the turning pink of the litmus paper does not tell Chris that there are no nonacids in his portion of the world capable of turning litmus paper pink. How could it? Litmus paper doesn't work that way. It just turns pink or not. If it turns pink and the channel condition holds, it informs Chris of the presence of an acid. Chris knows acids are not nonacids capable of turning litmus paper pink. So he is informed that p (acid in beaker). Knows $p \rightarrow q$ (if it is an acid, then it is not a nonacid capable of turning litmus paper pink). But does not know and is not informed that q (the liquid in the beaker is not a nonacid capable of turning litmus paper pink).

So how can Chris know it *is an acid*, you ask. Because he never once doubts the litmus test. It is indeed an acid. There are in fact no nonacids in Chris's world capable of turning litmus paper pink. The litmus test does indeed *inform Chris that p*. Chris believes, his belief is true, sustained by information, and is not Gettierizable. That is all that is required for knowledge on a reasonable information-based account of knowledge (or similar tracking theory). That's how Chris knows. Of course Chris believes the channel condition holds, or he would not trust the litmus test. But its holding can be in the background—that is, he need not have the information that it holds. He can believe it with justification—but the justification does not have to guarantee the truth of this belief about the channel condition. Indeed, as long as justified beliefs can be false, they will *not* guarantee the truth of his belief about the channel conditions. And further, Floridi himself has a supposed "proof" that three conditions alone (belief, truth, and justification) cannot block Gettierization. So, presumably, Floridi himself is *on board* with this account (even if others are not—as I presume will be the case—given the enormous acceptance of closure in spite of the reasons I am now giving to the contrary).[4]

I suspect the opposition to continue to press that if one lacks knowledge that it is not a nonacid capable of turning litmus paper pink, how can one have knowledge that it is an acid? So I shall try harder.[5] Because if in fact

---

[4] Whereas Dretske's and Nozick's accounts do handle these puzzles, it is not so clear how Floridi's weaker account of being informed can solve these puzzles.

[5] For more see Adams, Fugurelli, and Barker unpublished, and Barker and Adams unpublished.

there are no nonacids capable of tuning litmus paper pink in one's relevant environmental niche, one does not need a detector (another detector of *ph*) to eliminate that alternative (it is not a *relevant alternative*). Such a possibility is too remote to be relevant. However, to *knowingly claim* such possibilities are too remote to be relevant (or actual) requires a reliable detector of them—a detector one has not got in the case at hand.

To see this, suppose that what Chris detects—call it *x*—is in fact an acid, and that Chris knows *x* is an acid by means of a litmus detector D (*x*'s turning D pink). Suppose in addition that D is sensitive to *x*'s being an acid, but not to *x*'s not being a nonacid capable of turning litmus paper pink. How can D possibly enable one to know that x is an acid? If in fact there are no nonacids capable of turning litmus paper pink in one's relevant environmental niche, *x*'s being a nonacid capable of turning litmus paper pink is an *irrelevant alternative* to *x*'s being an acid. An alternative is a *relevant alternative* with respect to knowing that *x* is an acid if and only if the alternative might obtain were *x* not an acid. If there were what we might call *fake* acids *nearby*, and if *x* might be such a fake, were *x* not an acid, then *x*'s being a fake acid would be a relevant alternative. In such a case, Chris's believing that *x* is an acid on the basis of *x*'s turning the litmus paper pink would not qualify as knowledge, for one's litmus test would not function as an acid detector—were *x* not an acid, *x* might be a fake acid, and therefore might turn litmus paper pink. Assuming that the setting is a laboratory environment devoid of such fakes, however, *x*'s being a fake acid is not a relevant alternative, since *x* would not be such a fake were *x* not an acid. Instead *x* would be something that presents a significantly different result to the litmus test. In one's actual environment, *x*'s turning litmus paper pink *is* sensitive to *x*'s being an acid, and functions as a *acid detector* that enables one's belief to qualify as knowledge. Nevertheless, with respect to knowing *that x is an acid but not/rather than a fake acid,* *x*'s being such a fake qualifies as a relevant alternative, for regardless of environmental conditions—were it not the case that *x* is an acid rather than a fake acid, *x* would be a fake acid rather than an acid. Consequently, in order to know that *x* is an acid rather than a fake acid, one would need a detector (e.g., passing another chemical analysis) that would be sensitive both to *x*'s being an acid and to *x*'s not being a fake acid. Lacking this alternative test, Chris can know at most that *x* is an acid, but not that it is not a nonacid capable of turning litmus paper pink.

Hence, controversial as it may be, information-based tracking theorists of knowledge are in a position to resist at least this one piece of Floridi's axiomatization of an informational logic.[6] However, the rest of his account seems to me correct and a significant addition to formalization of doxastic and epistemic logics.

---

[6] In fact, see Arlo-Costa and Parikh 2006, where this is discussed in the context of modal logics.

Acknowledgments

I would especially like to thank Patrick Allo, Fred Dretske, and Luciano Floridi: Allo for very helpful advice editing and preparing this chapter, Dretske for teaching me the application of information to semantics over many years, and Floridi for developing and extending the application of information to many areas of philosophy.

References

Adams, Fred. 2003. "Thoughts and Their Contents: Naturalized Semantics." In *The Blackwell Guide to Philosophy of Mind*, edited by Ted Warfield and Stephen Stich, 143–71. Oxford: Basil Blackwell.

———. 2004. "Knowledge." In *The Blackwell Guide to the Philosophy of Information and Computing*, edited by Luciano Floridi, 228–36. Oxford: Basil Blackwell.

Adams, Fred. 2005. "Tracking Theories of Knowledge." *Veritas* 50:11–35.

———. Unpublished. "Consciousness: Why and Where?"

Adams, Fred, and Ken Aizawa. 2008. *The Bounds of Cognition*. Oxford: Blackwell.

Adams, Fred, and Steven Beighley. Forthcoming. "The Mark of the Mental." In *Mind Companion*, edited by James Garvey. London: Continuum.

Adams, Fred, John Barker, and Julia Fugurelli. Unpublished. "Towards Closure on Closure."

Adams, Fred, and Murray Clarke. 2005. "Resurrecting the Tracking Theories." *Australasian Journal of Philosophy* 83:207–21.

Aizawa, Ken, and Fred Adams. 2005. "Defending Non-Derived Content." *Philosophical Psychology* 18:661–69.

Arlo-Costa, Horacio, and Rohit Parikh. 2006. "Tracking Truth: Knowledge and Conditionals in the Context of Branching Time." Formal Epistemology Workshop.

Barker, John, and Fred Adams. Unpublished. "Knowledge, Action, and Closure."

Carruthers, Peter. 2005. *Consciousness*. Oxford: Clarendon Press.

Dretske, Fred. 1970. "Epistemic Operators." *Journal of Philosophy* 24:1007–23.

———. 1971. "Conclusive Reasons." *Australasian Journal of Philosophy* 49:1–22.

———. 1981. *Knowledge and the Flow of Information*. Cambridge, Mass.: MIT/Bradford.

———. 1988. *Explaining Behavior*. Cambridge, Mass.: MIT/Bradford.

———. 1995. *Naturalizing the Mind*. Cambridge, Mass.: MIT/Bradford.

———. 2005. "Is Knowledge Closed Under Known Entailment? The Case Against Closure." In *Contemporary Debates in Epistemology*, edited by Matthias Steup and Ernest Sosa, 13–26. Oxford: Blackwell.

Evans, Gareth. 1982. *Varieties of Reference*. Oxford: Oxford University Press.

Fitch, Tecumseh. 2008. "Nano-Intentionality: A Defense of Intrinsic Intentionality." *Biology and Philosophy* 23:157–77.

Floridi, Luciano. 2004a. "Open Problems in the Philosophy of Information." *Metaphilosophy* 35:554–82.

———. 2004b. "On the Logical Unsolvability of the Gettier Problem." *Synthese* 14:61–79.

———. 2004c. "Information Inspiration." *Philosopher's Magazine* 28, 4:56–60.

———. 2005. "Is Information Meaningful Data?" *Philosophy and Phenomenological Research* 70:351–70.

———. 2006. "The Logic of Being Informed." *Logique et Analyse* 196:433–60.

Floridi, Luciano, and Mariarosaria Taddeo. 2005. "The Symbol Grounding Problem: A Critical Review of Fifteen Years of Research." *Journal of Experimental and Theoretical Artificial Intelligence* 17:419–45.

———. 2007. "A Praxical Solution to the Symbol Grounding Problem." *Minds and Machines* 17:369–89.

Floridi, Luciano, Mariarosaria Taddeo, and Matteo Turrili. 2009. "Turing's Imitation Game: Still a Challenge for All Machines and Some Judges." *Minds and Machines* 19:145–50.

Fodor, Jerry. 1990. *A Theory of Content and Other Essays*. Cambridge, Mass.: MIT/Bradford.

Harnad, Stevan. 1990. "The Symbol Grounding Problem." *Physica D* 42:335–46.

Millikan, Ruth. 1989. "Biosemantics." *Journal of Philosophy* 86:281–97.

Nozick, Robert. 1981. *Philosophical Explanations*. Cambridge, Mass.: Harvard University Press.

Searle, John. 1983. *Intentionality*. Cambridge: Cambridge University Press.

Tye, Michael. 1995. *Ten Problems of Consciousness: A Representational Theory of the Phenomenal Mind*. Cambridge, Mass.: MIT/Bradford.

7

ABSTRACTION, LAW, AND FREEDOM IN COMPUTER SCIENCE

TIMOTHY COLBURN AND GARY SHUTE

Introduction

Despite its title, this chapter has nothing to do with legal, moral, or political issues. Had the title been "Abstraction, Law, and Freedom in *Cyberspace*," it no doubt would have had something to do with such issues, since it would have been about the behavior of people. But our concern is with computer science as a science, and since the subject matter of computer science is interaction patterns among various computational abstractions, we address the concept of "law" in the scientific, rather than the legal, sense.

Our topic bears on Luciano Floridi's work in philosophy of information in two ways. First, to understand the role of law in computer science, one must understand the world in which computer scientists work, a world consisting primarily of abstractions. Floridi, acknowledging a certain debt to computer science, has recently advocated for "the method of levels of abstraction" (Floridi 2008a) as an approach to philosophical investigation in general. Second, our investigation into the concept of law in computer science reveals a paradigmatic case of Floridi's "informational nature of reality" (Floridi 2008b). In characterizing reality informationally, Floridi appeals again, in part, to a computing discipline— namely, software engineering and its current focus on object-oriented programming. It is fruitful to consider, while we explicate our account of abstraction and law in computer science, whether it impacts the related but more general philosophical claims of Floridi.

As a science that deals essentially with abstractions, computer science *creates its own subject matter*. The programs, algorithms, data structures, and other objects of computer science are abstractions subject to logical— but not physical—constraints. We therefore expect its laws to be different in substantial ways from laws in the natural sciences. The laws of computer science go beyond the traditional concept of laws as merely *descriptive*, allowing *prescriptive* laws that constrain programmers while also describing in a traditional scientific lawlike way the behavior of physically realized computational processes.

Laws of nature describe phenomena in ways that are general and spatially and temporally unrestricted, allowing us to explain and predict specific events in time. Computer scientists also need to explain and predict specific events, but these events occur in computational worlds of their own making. What they need is not laws that describe the phenomena they bring about; instead they need laws that prescribe constraints for their subject matter, keeping it within limits of their abstract understanding, so that the concrete processes that ultimately run on physical machines can be controlled.

In the section "Computer Science as the Master of Its Domain," we describe the language and data abstraction employed by software developers and compare them to Floridi's concept of levels of abstraction. Floridi also borrows the technical concept of "objects" from the object-oriented programming of software engineering, and we consider how objects fit with a structural account of reality. In "The Concept of Law in Computer Science," we show how laws of computer science are prescriptive in nature but can have descriptive analogs in the physical sciences. We describe a law of conservation of information in network programming, and various laws of computational motion for programming in general. These latter laws often take the form of *invariants* in software design. In "Computer Science Laws as Invariants," we show how invariants can have both pedagogical and prescriptive effects by describing how invariants are enforced in a famous sorting algorithm conceived by C. A. R. Hoare.

A prescriptive law, as an invariant applied to a program, algorithm, or data structure, specifies constraints on objects in abstract computational worlds. Being abstract, computational worlds are products of programmers' creative imaginations, so any "laws" in these worlds are easily broken. For programmers, the breaking of a prescriptive law is tantamount to the relaxing of a constraint. Sometimes, bending the rules of their abstract reality facilitates algorithm design, as we demonstrate finally through the example of search trees in the section "The Interplay of Freedom and Constraint."

Computer Science as the Master of Its Domain

Computer science is distinct from both natural and social science in that *it creates its own subject matter*. Natural science has nature and social science has human behavior as subject matter, but in neither case is nature or human behavior actually *created* by science; the two are *studied*, and observations made through such study are *explained*. Computer science, however, at once creates and studies abstractions in the form of procedures, data types, active objects, and the virtual machines that manipulate them.

Computer science shares with mathematics the distinction of studying primarily abstractions. However, as we argue in Colburn and Shute 2007, the primary objective of mathematics is the creation and manipulation of *inference structures*, while the primary objective of computer science is the creation and manipulation of *interaction patterns*. Certainly, the objective of computer science is at times similar to that of mathematics—for example, when proving theorems about formal languages and the automata that process them. However, the central activity of computer science is the production of software, and this activity is primarily characterized not by the creation and exploitation of inference structures but by the modeling of interaction patterns. The kind of interaction involved depends upon the level of abstraction used to describe programs. At a basic level, software prescribes the interacting of a certain part of computer memory, namely, the program itself, and another part of memory, called the program data, through explicit instructions carried out by a processor. At a different level, software embodies algorithms that prescribe interactions among subroutines, which are cooperating pieces of programs. At a still different level, every software system is an interaction of computational processes. Today's extremely complex software is possible only through abstraction levels that obscure machine-oriented concepts. Still, these levels are used to describe interaction patterns, whether they be between software objects or between a user and a system.

What is a "level of abstraction" in computer science? The history of software development tells a story of an increasing distance between programmers and the machine-oriented entities that provide the foundation of their work, such as machine instructions, machine-oriented processes, and machine-oriented data types. Language abstraction accounts for this distance by allowing programmers to describe computational processes through linguistic constructs that hide details about the machine entities by allowing underlying software to handle those details. At the most basic physical level, a computer process is a series of changes in the state of a machine, where each state is described by the presence or absence of electrical charges in memory and processor elements. But programmers need not be directly concerned with machine states so described, because they can make use of languages that allow them to think in other terms. An assembly language programmer can ignore electrical charges and logic gates in favor of language involving *registers*, *memory locations*, and *subroutines*. A C language programmer can in turn ignore assembly language constructs in favor of language involving *variables*, *pointers*, *arrays*, *structures*, and *functions*. A Java language programmer can ignore some C language constructs by employing language involving *objects* and *methods*. The concepts introduced by each of these languages are not just old concepts with new names. They significantly enlarge the vocabulary of the programmer with new functionality while simultaneously freeing the programmer from having to attend to tedious details. For example, in the move from C to Java, programmers have new access to *active* objects, that is, data structures that are encapsulated with behavior so that they amount to a simulated network of software computers, while at the same time being released from the administrative work of managing the memory required to create these objects.

As levels of programming language abstraction increase, the languages become more expressive in the sense that programmers can manipulate direct analogs of objects that populate the world they are modeling, like shopping carts, chat rooms, and basketball teams. This expressiveness is only possible by hiding the complexity of the interaction patterns occurring at lower levels. It is no accident that these levels mirror the history of computer programming language development, for hiding low-level complexity can only be accomplished through a combination of more sophisticated language translators and runtime systems, along with faster processing hardware and more abundant storage.

We have shown that other forms of computer science abstraction besides language abstraction, for example procedural abstraction (Colburn 2003) and data abstraction (Colburn and Shute 2007), are also characterized by the hiding of details between levels of description, which is called *information hiding* in the parlance of computer science.

Floridi proposes to use levels of abstraction to moderate long-standing philosophical debates through a method that "clarifies assumptions, facilitates comparisons, enhances rigour and hence promotes the resolution of possible conceptual confusions" (Floridi 2008a, 326). These features are certainly advantages in a philosophical context, and there is no doubt that abstraction through information hiding in computer science goes a long way toward mitigating "conceptual confusions," but how similar are Floridi's and computer science's levels of abstraction provided by programming languages?

In Floridi's view, a level of abstraction is, strictly speaking, a collection (or "vector") of observables, or interpreted typed variables. What makes the collection an abstraction is what the variables' interpretations choose to ignore. For example, a level of abstraction (LoA) of interest to those purchasing wine may consist of the observables *maker*, *region*, *vintage*, *supplier*, *quantity*, and *price*, while a LoA for those tasting wine may consist of *nose*, *robe*, *color*, *acidity*, *fruit*, and *length* (2008a, 309). So a LoA by itself is, loosely, a point of view, one chosen simply to suit a particular purpose.

One can see how the observables available to descriptions of computational processes in different programming languages provide different views of the entities participating in computational models of the world. Some programmers see registers and subroutines, others see variables and functions, and still others see objects and methods. But while the choice of wine LoA between the two examples given above would be made solely on the basis of some underlying purpose, the choice of abstraction level for a software application involves considering both an underlying purpose and a need to make use of a higher LoA. While there are some exceptions in the world of specialized hardware programming, most software applications for general purpose machines today require enhanced expressiveness through LoAs that do not require the programmer to be concerned with the architecture of the machine. The level of abstraction used by one programming language is "higher" than another to the extent that it hides more machine architecture details and allows the natural representation of concepts in the world being modeled by a program.

Computer scientists often engage in comparisons of programming languages on the basis of their expressiveness or fitness for various purposes. If a programming language can be said to embody a single LoA, computer scientists would therefore be interested in the relationships of multiple languages through the arrangement of multiple LoAs, something Floridi considers in his concept of a "gradient" of abstractions (GoA). To develop the idea of a GoA, Floridi first introduces the notion of a "moderated" LoA. In general, not all possible combinations of values for variables in a LoA are possible—for example, wine cannot be both white and highly tannic. A predicate that constrains the acceptable values of a LoA's variables is called its "behavior." When you combine a LoA with a behavior, you have a moderated LoA (2008a, 310).

Before considering GoAs for programming languages, let us consider the concept of a programming language as a moderated LoA, that is, one with a "behavior." While there is an obvious sense in which some observables, such as program variables, have value constraints (a program variable, for example, cannot be both integral and boolean), other observables, such as functions, have no value constraints other than the syntax rules that govern their construction. Yet functions nevertheless possess behavior.

There is a ready explanation for this disconnect. The model for Floridi's behavior concept is inspired by the information modeling activity of system specifiers, whose objective is the complete functional description of particular envisaged applications in terms of system state changes. Programming language developers, however, are in the quite different business of providing the tools that facilitate the implementation of the products of the specifiers, and programmers themselves use those tools to create computational objects of varying degrees of expressivity. It seems possible in principle to extend Floridi's concept of behavior to cover function behavior as well as the behavior of typed variables.

Returning now to Floridi's idea of a gradient of abstractions, we find that a GoA is constructed from a set of moderated LoAs. The idea, according to Floridi, is that "[w]hilst a LoA formalises the scope or granularity of a single model, a GoA provides a way of varying the LoA in order to make observations at differing levels of abstraction" (2008a, 311). Why would one want to do this? By way of explanation, Floridi again considers the wine domain. Since tasting wine and purchasing wine use different LoAs, someone who is interested in both tasting and purchasing could combine these LoAs into one GoA that relates the observables in each. So while a LoA is characterized by the predicates defined on its observables, a GoA requires explicit relations between each pair of LoAs. Floridi describes these relations formally using standard Cartesian product notation and conditions that ensure that the related LoAs have behaviors that are consistent. As Floridi points out, these consistency conditions are rather weak and do not define any particularly interesting relations between the LoAs. However, by adding certain conditions he defines a "disjoint" GoA, or one whose pairwise LoAs have no observables in common, and a "nested" GoA, or one whose LoAs can be linearly arranged such that the only relations are between adjacent pairs (2008a, 312).

Nested GoAs are useful because they can "describe a complex system exactly at each level of abstraction and incrementally more accurately" (2008a, 313), as in neuroscientific studies that begin by focusing on brain area functions generally and then move to consideration of individual neurons. It is interesting to note that while the gradient in a nested GoA goes from more abstract to more concrete, the gradient at work in computer science programming language abstraction proceeds along an orthogonal dimension, namely, from the more machine-oriented to the more world-oriented.

Put another way, nested GoAs (though not GoAs in general) are often constructed for the sake of more fine-tuned *analysis*, while the new abstraction levels offered by computer science programming languages present greater opportunities for *synthesis* in the construction of programming objects of ever-larger content than those available before. The content of a programmer's computational objects *enlarge* as more of the underlying machine details of representing them are hidden from the programmer by the chosen implementation language. This is particularly evident when considering the data abstraction capabilities of higher-level languages.

Lower-level languages such as assembly languages offer basic data types like numbers, characters, and strings. These are types that are themselves abstractions for the hard-wired circuitry of a general-purpose computer. If a programmer wants to write, say, a Web browser using assembly language, he must laboriously implement a high-level notion such as a communication socket in the language of the machine— numbers, characters, strings, and so on. As all assembly programmers know, this is an impoverished vocabulary requiring painstaking and time-consuming coding that is unrelated to the central problems of Web browsing. Hundreds or perhaps thousands of lines of code might be required just to perform what amounts to machine-level data book-keeping. By using a higher-level language such as C, the programmer can take advantage of richer types such as structures (that can group the basic machine types automatically) and pointers (that can take on the burden of memory address calculations), relieving a good deal of the drudgery of manipulating these basic machine types. But by using higher-level languages still, such as C++ or Java, programmers can expand their coding language to include types whose values are communication sockets themselves. No "shoehorning" of the higher-level notion of a communication socket using lower-level programming types is necessary; a communication socket actually *is* a data type in these higher-level languages.

It may very well be possible to define a gradient of abstractions for programming languages and data types that would fit Floridi's model, so long as the relations between the LoAs making up the gradient accurately capture the nature of information hiding on which software development so crucially depends. In fact, the concept of a GoA may be useful for characterizing the relations among the various "paradigms" of programming languages. For example, the fundamentally imperative and machine-oriented nature of assembly language programming and the purely functional nature of Lisp or Scheme may place them in a disjoint GoA, while the historical evolution of the C language into C++ suggests a nested GoA for them. Floridi points out (2008a, 314) that hierarchical GoAs that are less restricted than nested GoAs can also be defined, arranging their LoAs in tree structures. The many programming paradigms that have evolved in computer science would likely fit on such a GoA.

Whatever the programming paradigm, by hiding the complexity of dealing with lower-level data types, the information hiding provided by high-level languages gives programmers and scientists the expressiveness and power to create higher-level objects that populate ever more complex worlds of their own making. Such worlds exhibit Floridi's informational structural realism. Structural realism (SR) gives primacy to the *relations* that hold among objects being studied, for it is they, rather than the *relata* themselves, that facilitate explanation, instrumentation, and prediction within the system under investigation (Floridi 2008b, 220). What makes Floridi's SR informational is that questions about the ontological status of the *relata* can be put aside in favor of a minimal, informational conception of the objects bearing the relations. To explicate this idea, Floridi borrows the technical concept of an "object" from computer science's discipline of *object-oriented programming* (OOP). Because such objects encapsulate both state (Floridi's observables in a LoA) and behavior (how they relate or interact with other objects), they constitute paradigm candidates for the structural objects embraced by SR. But because they are abstractions, they avoid any ontological calls for substance or material in the characterization of structural objects.

The ontologically noncommittal nature of Floridi's informational structural realism (ISR) comes through when he uses Van Fraassen's categorization of types of structuralism and puts ISR in the category of "in-between" structuralism: it is neither radical (structure is all there is) nor moderate (there are nonstructural features that science does not describe). Instead, "the structure described by science does have a bearer, but that bearer has no other features at all" (Floridi 2008b, 221f.). However, as any object-oriented programmer knows, information objects are richly featured, even apart from the important relations they bear to other objects. A file folder, for example, has a size and a location besides the relations to the parent folder that contains it and the subfolders it contains. True, the objects of OOP are structured, but they are not mere relations; they are rich with *attributes*—OOP parlance for nonrelational properties. A programmer could, of course, replace an object O's attributes with other objects, so that O consists only of relations with other objects, but eventually those other objects must have attributes that are nonrelational. So not all objects can be featureless, which Floridi seems to desire.

At the same time, one of the powerful features of OOP that distinguishes it sharply from the programming paradigms preceding it is that program objects can *be* relations themselves, not just participate in them. So a programmer modeling a ticket agency can describe *spectator* and *show* objects, but it is possible (and in most cases preferable) to model the relationship that is a spectator's attending a show as itself an object as well.

Such is the generic nature of the "object" in OOP, whether an object is a relation or a relatum is entirely dependent on context. The object-oriented programmer has not only expressive but also ontological freedom in crafting his objects, and with that freedom comes the need for constraints on object behavior in the form of laws.

The Concept of Law in Computer Science

Modern epistemology is concerned in part with how we use our sense experience to acquire immediate knowledge of individual objects, processes, and events in the physical world through the interaction of our own bodies with it. But as John Hospers remarks, "If our knowledge ended there, we would have no means of dealing effectively with the world. The kind of knowledge we acquire through the sciences begins only when we notice *regularities* in the course of events" (1967, 229). When a statement of such a regularity admits of no exceptions, as in *Water at the pressure found at sea level boils at 212 degrees Fahrenheit*, it is called a "law of nature" (230).

There is a rich twentieth-century analytic philosophy tradition of trying to characterize exactly what scientific "lawlikeness" is, but there is general agreement that, at the very least, a statement is lawlike if it is general (i.e., universally quantified), admits of no spatial or temporal restrictions, and is nonaccidental (i.e., is in some sense necessary) (see, e.g., Danto and Morgenbesser 1960, 177). When knowledge has these properties, it can be used to both explain and predict particular observed phenomena and occurrences. Being able to predict what will happen in a given situation allows us to control future events, with the attendant power that accompanies such control.

Computer science, as we mentioned, does not study nature, at least the nature studied by physics, chemistry, or biology. It studies, of course, *information* objects (in a general sense, not necessarily in the sense of OOP), and most software developers would view their reality in Floridi's way as "mind-independent and constituted by structural objects that are neither substantial nor material . . . but informational" (Floridi 2008b, 241). But if computer science is a science, albeit a science concerned with information objects, and science attempts to discover empirical laws, with what laws, if any, are computer scientists concerned?

Some pronouncements, for example, the celebrated *Moore's Law*, are empirically based predictions about the future of technology. Moore's Law makes the observation that the number of integrated circuit components (such as transistors) that can fit on a silicon wafer will double every two years. While this observation, a version of which was first made in 1965, has proved uncannily correct, it does not fit the criteria for a scientific law, since even those who uphold it do not believe that it is temporally unrestricted; the technology of integrated circuit creation has physical limits, and when these limits are reached Moore's "Law" will become false.

Other purported "laws" having to do with the future of technology have already proven false. *Grosch's Law*, also coined in 1965, stated that the cost of computing varies only with the square root of the increase in speed, and so it supported the development of large supercomputers. But the opposite has emerged: economies of scale in most cases are achieved by clustering large numbers of ordinary processors and disk drives.

If Moore's Law and Grosch's Law are technology reports, other statements seem to embody, if not laws of nature, then *laws of computation*. For example, Alan Turing demonstrated that no procedure can be written that can determine whether any given procedure will halt or execute indefinitely. Such a statement satisfies the lawlikeness criteria given above, namely, generality, spatiotemporal independence, and being nonaccidental, but it is a statement of a mathematical theorem, not an empirical law supported by observation.

Other pronouncements that come to be known as "laws" are also really mathematical relationships applied to problems in computer science. *Amdahl's Law*, for example, gives an upper bound on the speedup one can expect when attempting to parallelize a serial algorithm by running it concurrently on a fixed number of processors. Similarly, *Gustafson's Law* modifies Amdahl's Law by removing the constraint that the number of processors be fixed. Each of these laws is deduced a priori from abstract concepts, not a posteriori from observations.

There is another important set of mathematical laws with which computer scientists are fundamentally concerned, dealing with the bounds on space and time that are imposed on computational processes by their inherent complexity. Just as Amdahl's and Gustafson's laws use formal mathematical arguments to give limits on the speedup obtained through the use of multiple processors, computer scientists use mathematical methods to classify individual algorithms, for example algorithms to search various kinds of data structures, in terms of the memory required to store the data structures and the number of operations required to search them. It is important to note that these analyses are based not on actual physical runnings of the algorithms through programs running on physical machines but on the analysis of algorithms as abstract mathematical objects. No empirical study or investigation is involved; in fact, a typical objective of such analysis is to determine whether a given computation can be accomplished within reasonable space and time bounds regardless of technology prognostications regarding the speed of future machines. Such statements about abstract computational objects may seem to satisfy the above criteria of law-likeness, but these statements are supported by formal proofs and not empirical investigation.

So is there a sense in which computer science can be said to "discover" scientific laws in the empirical sense? We think computer scientists don't discover laws; they must make them. When a programmer specifies an abstract procedure for sorting an abstract list of objects using the operations of an abstract machine, it is because that procedure will be implemented, through a complex and remarkable chain of electronic events, on a given machine in the physical world. Those events must be able to be accurately predicted in order for a program to serve the purpose for which it was intended. For this to happen, these electronic events must obey laws, but to *describe* these laws at the physical level of electrons and microcircuits would serve no purpose, because the physical laws of electromagnetism at the molecular level are irrelevant to a programmer's objective, which is to describe, in the formal language of a machine, the interaction patterns of abstract objects like variables, procedures, and data structures. Instead, the electronic events unfold as they do, and (one hopes) as they should, because the programmer *prescribes* laws in the realm of the abstract. Programmers must make their abstract worlds behave as though there were laws, so that the physical processes they produce benefit from the explanatory and predictive power that accompanies laws. To this end, computer science relies heavily, if not essentially, on *metaphor*.

In Colburn and Shute 2008 we describe how metaphors can help computer scientists treat the objects in their abstract worlds scientifically. For example, although there is no natural "law of conservation of information," network programmers make things work as though there were one, designing error detection and correction algorithms to ensure that bits are not lost during transmission. Their conservation "law" relies upon a *flow* metaphor; although bits of information do not "flow" in the way that continuous fluids do, it helps immeasurably to "pretend" as though they do, because it allows network scientists to formulate precise mathematical conditions on information throughput and to design programs and devices that exploit them.

The flow metaphor is pervasive and finds its way into systems programming, as programmers find and plug "memory leaks" and fastidiously "flush" data buffers. But the flow metaphor is itself a special case of a more general metaphor of *motion* that is even more pervasive in computer science. Talk of motion in computer science is largely metaphorical, since when you look inside a running computer the only things moving are the cooling fan and disk drives (which are probably on the verge of becoming quaint anachronisms). Yet descriptions of the abstract worlds of computer scientists are replete with references to motion, from program jumps and exits, to exception throws and catches, to memory stores and retrievals, to control loops and branches. This is to be expected, of course, since the subject matter of computer science is *interaction* patterns.

But the "motion" occurring in the computer scientist's abstract world would be chaotic if not somehow constrained, and so we place limits that are familiar to us from the physical world. In the case of network programming, we borrow from physics to come up with a law of conservation of information. For programming in general we borrow from physical laws of motion to come up with laws of computation in the form of *invariants*.

Computer Science Laws as Invariants

Just as natural laws admit of no exceptions, when programmers prescribe laws for their abstract worlds they must make sure they admit of no variation. The mandate of a descriptive law of nature is to discover and describe what *is*. The mandate of a prescriptive law of computation is to legislate what *must hold*, invariably, while a computation takes place. Here are some examples of loosely stated prescriptive laws:

- *The parity bit must equal the exclusive-or of the remaining bits* (used in error checking code)

- *The array index must not be outside the range [0..n–1]* where *n* is the size of the array (to avoid out-of-bounds references)
- *The segments of an array must at all times stand in some given relationships to one another* (to sum or sort the array, as in the examples below)

These prescriptions state *constraints* that must hold in their respective computational worlds. When a constraint is maintained throughout a computational interaction, for example, during the calling of a procedure or an iteration of a loop, it becomes an *invariant*. Invariants do double duty. First, they function as pedagogical devices for explaining algorithms (see, e.g., Gries 1981), even for people with little prior programming experience. Second, they have the generality of laws in that they are specified for algorithms, which are abstractions, and as such embody computational laws governing *any* concrete processes that implement them.

It is important to distinguish between the role of an invariant in an abstract computational process and that of an *axiom* in a formal model. Invariants are things we want to *make* true, while axioms are *assumed* to be true unless the model shows they are inconsistent. But while we endeavor to make invariants true, in keeping with computer science as the master of its domain they are often broken. There are two basic reasons for this. First, invariants often operate at levels of abstraction that cannot be enforced at the programming level; that is, programming requires steps that temporarily violate invariants, with later steps restoring them. Second, an invariant might be violated to adapt a computational structure or algorithm for a different purpose. In what follows we describe both of these scenarios.

Complex computational processes can be brought about by enforcing complex invariants, which themselves can be created by accommodating simpler ones. For example, an invariant for summing the entries of an array (a contiguous list of data) can be formulated as a picture showing the sum of the entries considered so far and the part of the array whose entries have not yet been added to the sum:

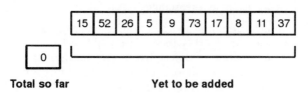

Total so far   **Yet to be added**

Here the partial sum has been initialized to zero and none of the entries has been added to it. The partial sum plus the entries not yet considered add up to the total of all entries in the array, which is 253. This is an invariant, because it remains true as the algorithm proceeds. After one entry is considered the picture looks like this:

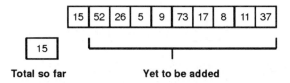

After two entries:

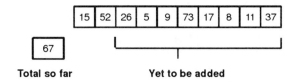

At each step the invariant holds. Eventually the partial sum is 253, and there are no more entries to consider. The invariant embodies a law that a computational process must follow in order to correctly total the entries of the array: it must manage the partial sum while reducing the unexamined array segment until it is empty.

When programming students are directed to implement this algorithm in a programming language, they learn that in order to keep track of the "Yet to be added" part of the array, they need to manage an index, call it i, that points to the beginning of that part. After having updated the partial sum with the first element of that part, the invariant is violated because the unexamined part has not been shrunk. It is up to the programmer to restore the invariant by incrementing i, which has the effect of shrinking the unexamined array segment. Describing invariants and how to maintain them through individual program steps is an essential part of the computer science education process. We shall see later that violating invariants (however temporarily) occurs not only during the programming process but also in the development of algorithms themselves.

The power of invariants can also be seen in a famous sorting algorithm developed by Hoare. To understand *Quicksort*, a sorting procedure celebrated for its efficiency, consider the same array as before:

$$\text{A} \quad \boxed{15\;52\;26\;5\;9\;73\;17\;8\;11\;37}$$

Hoare realized that if **A** could be partitioned into three segments, with all numbers in the left-most segment less than the single number (called the *pivot*) in the middle segment, and all numbers in the right-most segment greater than or equal to the pivot, as in this arrangement (15 is the pivot):

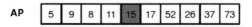

then the pivot element would be in its correct place in the final sorted array. Hoare further realized that if the left-most and right-most segments of **AP** were themselves partitioned in the same way, then *their* pivot elements would also be in their correct places in the final sorted array. If this process of partitioning segments is carried out recursively until each remaining segment has only zero or one element, the array will be completely sorted, and it will have been done efficiently provided that the partitioning was done efficiently.

The key to efficient partitioning of an array segment is to maintain three subsegments, which we will call <, ≥, and ?, that maintain the following invariant:

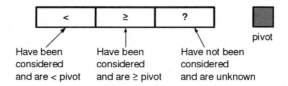

When the partitioning algorithm begins on **A**, the first element (15) is chosen as the pivot, all the non-pivot elements are unknown and part of the **?** segment, and the < and ≥ segments are empty. As the algorithm proceeds, the size of the unknown segment steadily decreases while < and ≥ grow, until finally the picture looks like:

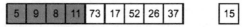

Here the < segment is shaded, the ≥ segment is unshaded, and the unknown segment **?** is empty. To finish the partitioning, the first element of the ≥ segment, 73, is replaced by the pivot and moved to the end of the array to produce the partitioned array **AP** shown above. The partitioning algorithm can now be recursively applied to the left and right subsegments to eventually sort the array.

It would have been difficult for anyone, even Hoare, to conceive of this algorithm without the help of a guiding invariant. Programmers start with invariants as prescriptive laws and then try to create abstract worlds that obey them. When an abstract process maintains a carefully pre-scribed invariant, its concrete realization will behave as though governed by a descriptive computational law. That is, its behavior will be predictable, controllable, and correct for its purpose. Thus invariants are to program objects what laws of nature are to natural objects. Just as a planet circling a sun cannot help but follow Kepler's laws of planetary motion and be predictable in its course, a program written to obey an invariant cannot help but behave in a predictable way.

The Interplay of Freedom and Constraint

Francis Bacon wrote, "Nature, to be commanded, must be obeyed" (Bacon 1889). This applies in computer science, but with a twist—program developers prescribe laws for programs, and then must ensure the programs obey these laws. The laws then become constraints on the programmer. But these are not constraints in a political sense. The danger of political lawlessness is external—your unconstrained freedom is a threat to my freedom and vice versa, but the danger of lawlessness in computer science is internal—our minds need constraints in order to reason through the consequences of programming decisions. For example, it is the programmer's imperative to reestablish the invariant for each iteration of a loop.

While constraints in general limit freedom, in the case of programming they make it much easier to achieve an objective, through the breaking down of a problem solution into a sequence of small steps, each governed by an invariant.

What seems to limit freedom actually opens up new pathways. Imagine a person exploring a jungle without a compass. She has freedom to move anywhere she wishes, but she has no guidance. As she makes her way, she has to frequently avoid entanglements, causing her to lose her direction. After a while, her path through the jungle is essentially a free but random walk. She has freedom of movement but lacks knowledge, so she cannot make meaningful progress.

Now give her a compass that constrains her movement but provides knowledge in the form of directional feedback. Even with entanglements she can proceed in a relatively straight line by correcting her direction after each deviation. By limiting her freedom in the short term (adding a compass correction after each entanglement), she increases her long-term freedom—her ability either to explore deeper into the jungle or to emerge from it.

The compass constrains the action of the jungle explorer just as program invariants constrain the action of computational processes. But in actual practice, programmers may or may not "think up" invariants before writing their programs. Beginning programmers often have a loose and vague idea of what their programs should do and start programming without constructing invariants. Just as a jungle explorer may get lucky and emerge without a compass, a programmer who disregards invariants may get lucky and produce a program whose behavior seems correct, but he cannot be sure that it is correct in all instances. Through invariants, however, programmers can be confident in their problem solutions, because the interactions they produce are governed by law, albeit prescriptive law.

While we have shown that programmers learn to manage arrays by temporarily violating and then restoring invariants, this approach can also be fruitful in algorithm development, a nonprogramming endeavor. For example, consider a data structure known as a *binary search tree* (BST). BSTs facilitate looking up data using a key. Their functionality is similar to telephone directories, where the keys are people's names and the data are addresses and telephone numbers. Here is a BST whose keys are simple strings (for simplicity, the data are not shown):

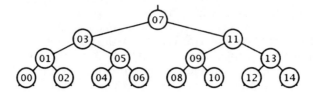

In order to preserve order among a BST's data items, a programmer must maintain the following invariant: for any node in the BST, all keys in its left subtree must be smaller than its key, and all keys in its right subtree must be greater than its key. When a new node is added to a BST, its key is compared with the key of the tree's root (the topmost node). If it is smaller, the new node will be placed in the left subtree, otherwise in the right. The appropriate subtree is recursively searched until an available space is found on one of the tree's leaves (bottommost nodes). Here is the example tree after the node with key **09a** is added:

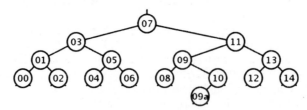

This arrangement facilitates data retrieval by key, since a key can be located in time proportional to the height of the tree. If a tree is balanced, as in the one above, a key can be located efficiently even if the number of nodes is large. For example, a balanced tree of one million nodes has a height of about 20.

Unfortunately, the structure of a BST is determined by the order in which nodes are added to it, so nothing guarantees that a BST will be balanced. Here is a BST in which nodes with keys **00**, **01**, **02**, **03**, **04**, **05**, and **06** have been added in that order:

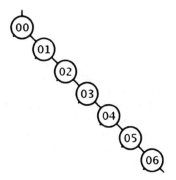

Although this structure satisfies the order invariant for BSTs, it cannot be efficiently searched, since it is not balanced. If one million nodes are added to a BST in key order, finding nodes with higher numbered keys will take time proportional to its height, which is one million (compared to 20 in a balanced BST of a million nodes).

To solve this problem, computer scientists have devised a kind of self-balancing BST known as a *red-black tree* (RBT). In addition to the ordering invariant imposed on BSTs, RBTs introduce the concept of a node's *color*, requiring every node to be either red or black, with the following additional constraints:

1. All downward paths from the top (root) of the tree to the bottom (leaves) must contain the same number of black nodes.
2. The parent of a red node, if it exists, is black.

Here is a BST that is also a RBT (red nodes are shown here as dotted circles):

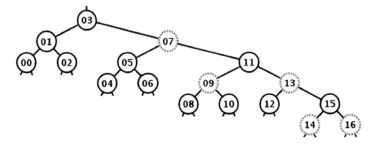

The RBT constraints do not make a search tree perfectly balanced, but they do ensure that no search path is longer than twice the length of search paths in a perfectly balanced tree. Thus in a tree with one million nodes, search paths will be no longer than 40.

This arrangement works without a hitch for some cases. Consider adding a node with key **09a**. Using the standard BST adding algorithm, it can be added as the left subtree of node **10**. Then it can satisfy the RBT constraints by being colored red. Now consider adding a node with key **17**. The BST adding algorithm would put it as the right subtree of node **16**. However, coloring it black would violate RBT constraint 1, while coloring it red would violate RBT constraint 2.

But because computer science is the master of its domain, we can choose to violate our own laws through the temporary relaxing of constraints. While the suspension of a natural law would be called a miracle, the temporary violation of our RBT constraints, with a little ingenuity, can result in an efficient data structure. So we go ahead and violate constraint 2 by adding **17** and coloring it red (shown here as a dotted circle):

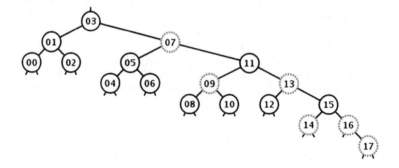

Without going into detail, suffice it to say that by tweaking this law-breaking RBT in various ways (through structural changes known as rotations and certain recolorings) it can be nudged back into being a law-abiding citizen of the computational world with both RBT constraints satisfied:

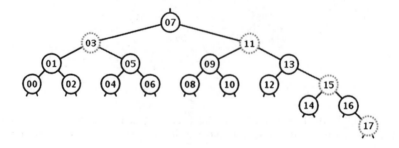

Conclusion

The ultimate nature of computational reality is, of course, informational. That is what allows computer science to create its own empirical laws, and to be free to break them when it sees fit. We have shown that the worlds of computational objects need laws in the form of self-prescribed invariants, but also that the suspension of these laws might be creative acts, resulting in neither anarchy nor miracles.

Whether the ultimate nature of *all* reality is informational is not as obvious. But if computer science has sage advice for philosophy, Floridi has seized upon the right concepts in levels of abstraction and the methodology of object-oriented programming, for they are the catalysts of progress in computer science and software engineering. While OOP may one day be replaced by another latest and greatest programming paradigm, the march toward ever-higher levels of programming abstractions will continue. Whether this march coincides with a move toward informational ontology is yet to be seen.

References

Bacon, Francis. 1889. *Novum Organum*. Edited by Thomas Fowler. Oxford: Clarendon Press.

Colburn, Timothy. 2003. "Methodology of Computer Science." In *The Blackwell Guide to the Philosophy of Computing and Information*, edited by Luciano Floridi, 318–26. Oxford: Blackwell.

Colburn, Timothy, and Gary Shute. 2007. "Abstraction in Computer Science." *Minds and Machines* 17:169–84.

———. 2008. "Metaphor in Computer Science." *Journal of Applied Logic* 6:526–33.

Danto, Arthur, and Sidney Morgenbesser. 1960. "Introduction to Part 2, Law and Theories." In *Philosophy of Science*, edited by Arthur Danto and Sidney Morgenbesser, 177–81. Cleveland, Ohio: Meridian Books.

Floridi, Luciano. 2008a. "The Method of Levels of Abstraction." *Minds and Machines* 18:303–29.

———. 2008b. "A Defence of Informational Structural Realism." *Synthese* 161:219–53.

Gries, David. 1981. *The Science of Programming*. New York: Springer.

Hoare, C. A. R. 1962. "Quicksort." *Computer Journal* 5:10–15.

Hospers, John. 1967. *An Introduction to Philosophical Analysis*. 2nd ed. Englewood Cliffs, N.J.: Prentice Hall.

8

STRUCTURALISM AND INFORMATION

OTÁVIO BUENO

Luciano Floridi has developed an insightful and sophisticated version of realism about structures: informational structural realism (Floridi 2008). One of the significant benefits of Floridi's proposal is the fact that it explicitly invokes information as a central component of structuralism. Information is arguably a central feature of science. Scientists typically search for informative theories; they develop informative ways of probing domains of inquiry by devising experiments about such domains, and they create instruments that offer information about otherwise inaccessible phenomena. Scientists also try to identify significant structural components of the phenomena they study. These structural components are variously represented by the relevant theories, experiments, and instruments' outputs. Thus, central to a philosophical account of science—particularly one that emphasizes structures as part of scientific practice—is to establish the connections between structure and information. To suggest one way such connections can be determined in light of Floridi's account of informational structuralism is the main goal of this chapter.

I will combine the partial structures framework (da Costa and French 2003; Bueno 1997, 2000) with the emphasis on information that animates so many aspects of Floridi's work. The result will be a form of structuralism that shares with Floridi's an emphasis on information and shares with the partial structures approach the capacity to accommodate the partiality of information that is so ubiquitous in scientific practice.

Information and Structural Realism

"Information" is *typically* a success term. If something is information for some point, it is *usually* taken to be true. The qualifiers, however, are important here. After all, there are clear cases in which there is information for some point even though the information in question is not true.[1]

---

[1] Floridi disagrees, given that, on his view, false semantic information is pseudo-information: it is not semantic information at all (Floridi 2005). As will become clear, a more nuanced account is in order.

Consider, for instance, the discovery of Neptune (Grosser 1979). In the nineteenth century, it became clear that the predicted orbit of Uranus (the last known planet in the solar system at the time), as described by Newtonian mechanics, did not match the observed record. Instead of simply taking this instance as a falsification of Newtonian physics, astronomers postulated the existence of an undetected planet that was interfering with the orbit of Uranus. In order to locate the planet, Bode's law was used. According to this law, the mean distance of each planet in the solar system from the Sun in astronomical units (AU) follows a determined pattern. By employing the pattern described by Bode's law, together with calculations from Newtonian physics, astronomers were able to identify the planet that was interfering with the orbit of Uranus. The new planet was called "Neptune." Bode's law and Newtonian physics were clearly crucial for the discovery of this planet.

But it turns out that Neptune showed Bode's law to be false. Neptune's mean distance from the Sun (in astronomical units) is 30.07 AU rather than the 38.80 AU predicted by Bode's law—a significant mismatch, well beyond any acceptable margins of experimental error. As we now know, Newtonian mechanics is also false. However, clearly both Bode's law and Newtonian physics provided information—in fact, key information—for the discovery of Neptune. Information can be used successfully, but it need not be true for it to play a successful role. Truth is not required for empirical success, not even novel empirical success involved in the discovery of a new planet.

Once this point is recognized, it becomes clear that informational content and truth content are quite distinct. A false theory, such as Newtonian mechanics, can be extremely informative. A true statement, such as a tautology from classical logic, can be completely uninformative. Of course, information is also contextual. A classical tautology can be quite informative in the context of a formal proof (e.g., it may help to shorten a given proof), and Newtonian mechanics can be significantly uninformative in the context of pediatrics (it will not help us to figure out how to treat cases of smallpox).

Is there a form of structuralism in philosophy of science for which informational content is crucial? Floridi's informational structural realism (Floridi 2008), with its sophisticated representation of information and the incorporation of suitable levels of abstraction, provides such a view. Informational structural realism is a form of realism in the basic sense that it is committed to the existence of a mind-independent reality. More important, it also defends the view that reality has structural properties and structural objects, and the objects are characterized in informational terms. The central features of structural objects are illustrated in the conceptualization of objects in chess. "Consider," writes Floridi,

> a pawn in a chess game. Its identity is not determined by its contingent properties as a physical body, including its shape and colour. Rather, a pawn is

a well-defined cluster of specific *states* (properties like white or black, and its strategic position on the board) and determined *behavioural rules* (it can move forward only one square at a time, but with the option of two squares on the first move . . .), which in turn are possible only in relation to other pieces and the logical space of the board. For a player, the actual pawn is only a placeholder, whereas the real pawn is an 'informational object'. It is not a material thing but a set of typed variables, using the LoA [level of abstraction] terminology, or a mental entity, to put it in Berkeley's terms, or an entity constituted by a bundle of properties, to use a Humean expression. Its existence and nature is determined by the differences and nomological relations that characterize the game of chess. The physical placeholder can be replaced by a cork without any semantic loss at the LoA required by the game. Indeed, a player may not even need a physical placeholder at all. (2008, 239)

This example highlights the crucial components of structural objects. Their particular physical constitution is not their defining feature. What is crucial is the sort of information structural objects encode given the specific states they can be in, and the behavioral rules they satisfy. And both states and rules are specified only in relation to other (structural) objects. The particular nature of such objects is not relevant.

With these considerations in place, the key feature of informational structural realism can be stated: "Explanatorily, instrumentally and predictively successful models (especially, but not only, those propounded by scientific theories) at a given LoA can be, in the best circumstances, increasingly informative about the relations that obtain between the (possibly sub-observable) informational objects that constitute the system under investigation (through the observable phenomena)" (Floridi 2008, 240–41). It is then clear that both informational content and informational objects (that is, structural objects characterized in informational terms) are fundamental to this form of structural realism.

However, high informational content is clearly not enough to guarantee realism—even at the level of structure. After all, despite its informational content, a scientific theory can still fail to be true, and in particular, even the information the theory provides about the structural features of a physical system can be false. A realist commitment about such structural features should then be withdrawn. Consider again the case of Newtonian mechanics. Given a certain level of abstraction, the theory clearly characterizes a suitable structure, which is then attributed to a corresponding physical system (see the diagram in Floridi 2008, 231, fig. 3). Although Newtonian physics enjoyed explanatory, predictive, and instrumental success, the structure that it describes—certain relations among point particles—is strictly false as a description of the physical system that characterizes the actual world. According to the Newtonian system, gravity is taken to be a force, mass does not depend on velocity, and there is no limit to the speed of light. These properties of the structure characterized by Newtonian mechanics do not correctly describe our actual physical system. Even without properly characterizing the structure of that system, Newtonian mechanics was unquestionably effective and informative. A form of realism that draws ontological conclusions from empirical success—even about structures rather than objects—runs into trouble here.

It may be argued that realists should not draw ontological conclusions from Newtonian mechanics, given that the theory is false. But similar problems emerge in the context of contemporary physics. Given the inconsistency between relativity theory and quantum mechanics, a serious issue emerges as to which of these theories (if any) offer the correct account of the structural features of the world, that is, which of these theories (if any) properly characterize a structure that successfully describes the physical system that constitutes the world. Perhaps only parts of such structures are successful in their descriptions—only parts of them successfully describe corresponding parts of the relevant physical systems. Perhaps a more local form of realism may be more promising than a global account that aims to describe the overall the structure of the world. If this is the case, we already have a significant concession, given that realism now becomes a far more limited claim (restricted to some aspects of the physical systems under consideration). And, of course, such partiality needs to be properly formulated. (I return to this issue in the next section.)

But let us grant that the commitments of informational structural realism have been adequately expressed. Floridi argues that informational structural realism offers a broad framework that reconciles epistemic structural realism and ontic structural realism (2008, 231–33). According to the epistemic structural realist, all we can know about the world is structure; according to the ontic structural realist, all there is in the world is structure (Ladyman 1998). The former is an epistemological claim, the latter an ontological view. On Floridi's reconstruction, the two views are perfectly compatible given that they operate at different levels of abstraction. Whereas epistemic structural realism runs at the level of first-order ontological commitments, insisting that relational structures are knowable, ontic structural realism navigates at the level of second-order ontological commitments, characterizing relata as structural objects. Informational structural realism steps in at precisely this point, offering an informational account of such structural objects, and reconciling both views along the way.

The worry here is whether such reconciliation is possible while still preserving realism. How exactly can an informational account of structural objects, which emphasizes the nonmaterial features of objects, be reconciled with the concrete nature of the objects in physics (the objects that supposedly characterize actual physical systems)? If informational structural realism is to be a form of realism about the physical world, and if structural objects are components of physical systems, then such objects had better not be abstract. Otherwise, it would be unclear how we could obtain the concrete physical world out of abstract structural objects. Floridi addresses a related concern (2008, 247). But a difficulty still seems to remain: the "plasticity" of structural objects, which are characterized in informational terms, prevents us from being able to draw ontological conclusions about what they ultimately are (besides being abstract). What counts for the characterization of structural objects are the sorts of information they provide regarding the relevant states and behavioral rules—and this leaves entirely open the issue of the nature of the objects in question.

It may be argued, in response, that this should not be a problem. After all, we are dealing with a structuralist position, according to which the ultimate nature of objects is unknowable (at least in one form of structural realism). However, structural objects are supposed to be the fundamental constituents of structural realism. If their nature turns out to be unknowable, then the status of the structures they yield is similarly unknowable. At this point, it seems that we are approaching a form of skepticism rather than realism.

Floridi also argues that, once we take into account suitable levels of abstraction, there is no room for a position that in its ontological commitments is weaker than epistemological structural realism and stronger than instrumentalism, which has minimal ontological commitments (Floridi 2008, 232). However, it seems that there is room for such a view, which falls short of being a form of realism, even though it emphasizes the significance of structure, but it is still stronger than instrumentalism. I think structural empiricism is a view that satisfies such requirements (Bueno 1999).[2]

According to structural empiricism, all that science allows us to know about the world is structure, but such structure is restricted to observable parts—the ultimate nature of the underlying unobservable structures cannot be known. As opposed to instrumentalism, scientific theories do have a truth-value, even though we are not able to know what the truth-values are, due to underdetermination considerations (see van Fraassen 1980). Moreover, scientific theories also provide understanding: of how the world could be if such theories were true (see van Fraassen 1991). In

---

[2] I also think that Bas van Fraassen's empiricist structuralism (see van Fraassen 2008) meets the relevant requirements. There is much in common between the two views (empiricist structuralism and structural empiricism). Both are forms of empiricism that emphasize the role played by structures in scientific practice. Both articulate antirealist views that take scientific theories literally, despite their emphasis on the point that truth need not be a norm for scientific activity. Structural empiricism is articulated in terms of partial structures and partial truth (see da Costa and French 2003; Bueno 1997), whereas empiricist structuralism does not presuppose such a formal framework. Despite the differences between these views, both seem to offer stronger forms of antirealism that go beyond instrumentalism while still avoiding the commitments found in epistemic structural realism.

contrast to instrumentalism, theories are more than simple instruments of prediction. They have a cognitive role. And this cognitive role is connected with the information such theories supply, in particular, information about the observable aspects of the world, and about possible ways of interpreting what is going on beyond the phenomena. How can structure and information be connected on this view?

Partial Information and Partial Structures

A crucial feature of scientific practice is the partiality of information that it deals with. This partiality can be accommodated formally in terms of the partial structures approach (see da Costa and French 2003; Bueno, French, and Ladyman 2002; Bueno 1997). This approach relies on three main concepts: partial relation, partial structure, and quasi-truth. One of the main motivations for introducing this proposal derives from the need to supply a formal framework in which the openness and incompleteness of the information that is dealt with in scientific practice can be accommodated. This is accomplished, first, by extending the usual notion of structure, in order to accommodate the partialness of information we have about a certain domain (introducing then the notion of a partial structure). Second, the Tarskian characterization of the concept of truth is generalized for partial contexts, which then leads to the introduction of the corresponding concept of quasi-truth.

The first step, then, to characterize partial structures is to formulate a suitable concept of a partial relation. In order to investigate a certain domain of knowledge Δ (say, the physics of particles), researchers formulate a conceptual framework that helps them systematize and interpret the information they obtain about Δ. This domain can be represented by a set D of objects (which includes *real* objects, such as configurations in a Wilson chamber and spectral lines, and *ideal* objects, such as quarks). D is studied by the examination of the relations that hold among its elements. However, it often happens that, given a relation R defined over D, we do not know whether all objects of D (or n-tuples thereof) are related by R, or we need to ignore some of the relations that are known to hold among objects of D, in order to study other relations about that domain in a tractable way. This is part of the incompleteness and partiality of our information about Δ, and is formally accommodated by the concept of a partial relation. The latter can be characterized as follows. Let D be a nonempty set. An n-place *partial relation* R over D is a triple $\langle R_1, R_2, R_3 \rangle$, where R_1, R_2, and R_3 are mutually disjoint sets, with $R_1 \cup R_2 \cup R_3 = D^n$, and such that: R_1 is the set of n-tuples that (we know that) belong to R; R_2 is the set of n-tuples that (we know that) do not belong to R, and R_3 is the set of n-tuples for which it is not known (or, for reasons of simplification, it is ignored that it is known) whether they belong or not to R. (Notice that if R_3 is empty, R is a usual n-place relation that can be identified with R_1.)

But in order to accommodate the information about the domain under study, a concept of structure is needed. The following characterization, spelled out in terms of partial relations and based on the standard concept of structure, offers a concept that is broad enough to accommodate the partiality usually found in scientific practice. A *partial structure A* is an ordered pair $\langle D,R_i \rangle_{i \in I}$, where D is a nonempty set, and $(R_i)_{i \in I}$ is a family of partial relations defined over D.[3]

We have now defined two of the three basic concepts of the partial structures approach. In order to spell out the last one (quasi-truth), we will need an auxiliary notion. The idea here is to use the resources supplied by Tarski's definition of truth. But since this definition is only for full structures, we have to introduce an intermediary notion of structure to link partial to full structures. This is the first role of those structures that extend a partial structure A into a full, total structure (which are called A-normal structures). Their second role is model-theoretic, namely, to put forward an interpretation of a given language and to characterize semantic notions. Let $A = \langle D,R_i \rangle_{i \in I}$ be a partial structure. We say that the structure $B = \langle D', R_i' \rangle_{i \in I}$ is an A-*normal structure* if (i) $D = D'$, (ii) every constant of the language in question is interpreted by the same object both in A and in B, and (iii) R_i' extends the corresponding relation R_i (in the sense that, each R_i', supposed of arity n, is defined for all n-tuples of elements of D'). Note that, although each R_i' is *defined* for all n-tuples over D', it holds for some of them (the R_{i1}'-component of R_i'), and it doesn't hold for others (the R_{i2}'-component).

As a result, given a partial structure A, there are several A-normal structures. Suppose that, for a given n-place partial relation R_i, we don't know whether $R_i a_1 \ldots a_n$ holds or not. One of the ways of extending R_i into a full R_i' relation is to look for information to establish that it *does* hold; another way is to look for contrary information. Both are prima facie possible ways of extending the partiality of R_i. But the same indeterminacy may be found with other objects of the domain, distinct from $a_1, \ldots, a_n$ (for instance, does $R_i b_1 \ldots b_n$ hold?), and with other relations distinct from R_i (for example, is $R_j b_1 \ldots b_n$ the case, with $j \neq i$?). In this sense, there are *too many* possible extensions of the partial relations that constitute A. Therefore we need to provide constraints to restrict the acceptable extensions of A.

In order to do that, we need first to formulate a further auxiliary notion (see Mikenberg, da Costa, and Chuaqui 1986). A *pragmatic*

---

[3] The partiality of partial relations and structures is due to the incompleteness of our knowledge about the domain under investigation. With additional information, a partial relation can become a full relation. Thus, the partialness examined here is not ontological but epistemic.

structure is a partial structure to which a third component has been added: a set of accepted sentences P, which represents the accepted information about the structure's domain. (Depending on the interpretation of science that is adopted, different kinds of sentences are to be introduced in P: realists will typically include laws and theories, whereas empiricists will add mainly certain regularities and observational statements about the domain in question.) A *pragmatic structure* is then a triple $A = \langle D, R_i, P \rangle_{i \in I}$, where D is a nonempty set, $(R_i)_{i \in I}$ is a family of partial relations defined over D, and P is a set of accepted sentences. The idea is that P introduces constraints on the ways that a partial structure can be extended (the sentences of P hold in the A-normal extensions of the partial structure A).

Our problem is: given a *pragmatic* structure A, what are the necessary and sufficient conditions for the existence of A-normal structures? Here is one of these conditions (Mikenberg, da Costa, and Chuaqui 1986). Let $A = \langle D, R_i, P \rangle_{i \in I}$ be a pragmatic structure. For each partial relation R_i, we construct a set M_i of atomic sentences and negations of atomic sentences, such that the former correspond to the n-tuples that satisfy R_i, and the latter to those n-tuples that do not satisfy R_i. Let M be $\cup_{i \in I} M_i$. Therefore, a pragmatic structure A admits an A-normal structure if and only if the set $M \cup P$ is *consistent*.

Assuming that such conditions are met, we can now formulate the concept of quasi-truth. A sentence α is *quasi-true* in a pragmatic structure $A = \langle D, R_i, P \rangle_{i \in I}$ if there is an A-normal structure $B = \langle D', R_i' \rangle_{i \in I}$ such that α is true in B (in the Tarskian sense). If α is not quasi-true in A, we say that α is *quasi-false* in A. Moreover, we say that a sentence α is *quasi-true* if there is a pragmatic structure A and a corresponding A-normal structure B such that α is true in B (according to Tarski's account). Otherwise, α is *quasi-false*.

The idea, intuitively speaking, is that a quasi-true sentence α does not describe, in a thorough way, the whole domain that it is concerned with; it describes only an aspect of it: the one that is delimited by the relevant partial structure A. After all, there are several different ways in which A can be extended to a full structure, and in some of these extensions α may not be true. Thus, the concept of quasi-truth is strictly weaker than truth: although every true sentence is (trivially) quasi-true, a quasi-true sentence may not be true (since it may well be false in certain extensions of A).

It may be argued that because quasi-truth has been defined in terms of full structures and the standard notion of truth, there is no gain with its introduction. But there are several reasons why this is *not* the case. First, as was just seen, despite the use of full structures, quasi-truth is weaker than truth: a sentence that is quasi-true in a particular domain—that is, with respect to a given partial structure A—may not be true if considered in an extended domain. Thus, we have here a sort of underdetermination—

involving distinct ways of extending the same partial structure—that makes the concept of quasi-truth especially appropriate for empiricists. Second, one of the points of introducing the concept of quasi-truth, as da Costa and French (2003) have argued in detail, is that in terms of this notion, a formal framework can be advanced to accommodate the openness and partialness that is typically found in science. Bluntly put, the actual information at our disposal about a certain domain is captured by a *partial* (but not full) structure A. Full, A-normal structures represent ways of extending the actual information which are possible according to A. In this respect, the use of full structures is a semantic expedient of the framework (in order to provide a definition of quasi-truth), but no epistemic import is assigned to them. Third, full structures can ultimately be dispensed with in the formulation of quasi-truth, since the latter can be characterized in a different way, though still preserving all its features, independently of the standard Tarskian type account of truth (Bueno and de Souza 1996). This provides, of course, the strongest argument for the dispensability of full structures (as well as of the Tarskian account) vis-à-vis quasi-truth. Therefore, full, A-normal structures are entirely inessential; their use here is only a convenient device.

To illustrate the use of quasi-truth, let us consider an example. As is well known, Newtonian mechanics is appropriate to explain the behavior of bodies under certain conditions (say, bodies that roughly speaking have a low velocity with respect to the speed of light, that are not subject to strong gravitational fields, and so on). But with the formulation of special relativity, we know that if these conditions are not satisfied, Newtonian mechanics is false. In this sense, these conditions specify a family of partial relations, which delimit the context in which Newtonian theory holds. Although Newtonian mechanics is not true (and we know under what conditions it is false), it is *quasi-true*; that is, it is true in a given context, determined by a pragmatic structure and a corresponding A-normal one (see da Costa and French 2003).

But what is the *relationship* between the various partial structures articulated in a given domain? Since we are dealing with partial structures, a second level of partiality emerges: we can only establish *partial* relationships between the (partial) structures at our disposal. This means that the usual requirement of introducing an isomorphism between theoretical and empirical structures (see van Fraassen 1980, 64) can hardly be met. After all, researchers typically lack full information about the domains they study. Thus, relations weaker than full isomorphism (and full homomorphism) need to be introduced (French and Ladyman 1997; French and Ladyman 1999; Bueno 1997).

In terms of the partial structures approach, however, appropriate characterizations of *partial isomorphism* and *partial homomorphism* can be offered (see French and Ladyman 1999; Bueno 1997; Bueno, French, and Ladyman 2002). And given that these notions are more open-ended than the standard ones, they accommodate better the partiality of structures found in scientific practice.

Let $S = \langle D, R_i \rangle_{i \in I}$ and $S' = \langle D', R'_i \rangle_{i \in I}$ be partial structures. So, each R_i is a partial relation of the form $\langle R_1, R_2, R_3 \rangle$, and each R'_i a partial relation of the form $\langle R'_1, R'_2, R'_3 \rangle$.[4]

We say that a partial function[5] $f: D \to D'$ is a *partial isomorphism* between S and S' if (i) f is bijective, and (ii) for every x and $y \in D$, $R_1 xy \leftrightarrow R'_1 f(x)f(y)$ and $R_2 xy \leftrightarrow R'_2 f(x)f(y)$. So, when R_3 and R'_3 are empty (that is, when we are considering total structures), we have the standard notion of isomorphism.

Moreover, we say that a partial function $f: D \to D'$ is a *partial homomorphism* from S to S' if for every x and every y in D, $R_1 xy \to R'_1 f(x)f(y)$ and $R_2 xy \to R'_2 f(x)f(y)$. Again, if R_3 and R'_3 are empty, we obtain the standard notion of homomorphism as a particular case.

There are two crucial differences between partial isomorphism and partial homomorphism. First, a partial homomorphism does not require that the domains D and D' of the partial structures under study have the same cardinality. Second, a partial homomorphism does not map the relation R'_i into a corresponding relation R_i. Clearly a partial homomorphism establishes a much less strict relationship between partial structures.

Partial isomorphism and partial homomorphism offer mappings among partial structures that are less tight than their corresponding full counterparts—isomorphism and homomorphism. Partial mappings, as transformations that connect different partial models that may be used in scientific practice, allow for the transferring of information from one domain into another—even when the information in question is incomplete. After all, if a sentence is quasi-true in a given partial structure S, it will also be quasi-true in any partial structure that is partially isomorphic to S (see Bueno 2000).

As a result, partial mappings can be used as mechanisms of representation in scientific practice. It is in virtue of the fact that certain models of a given phenomenon share some of the structure of the latter—in the sense that there is a partial mapping between the two—that these models can be used to represent the relevant features of the phenomenon under study. Of course, which features are relevant is a pragmatic matter, largely dependent on the context under consideration.

It is also possible to accommodate the significance of models in reasoning about the phenomena, even when only partial information is available about the objects under study. The partial information is encoded in a partial structure that represents selected aspects of the

---

[4] For simplicity, I'll take the partial relations in the definitions that follow to be two-place relations. The definitions, of course, hold for any n-place relations.

[5] A partial function is a function that is not defined for every object in its domain.

phenomena. Representation is always made from a particular perspective; it is, thus, selective. In other words, to represent is, ultimately, to select intentionally certain aspects of the target to stand for corresponding aspects of the source. In the case of scientific representation, models, broadly understood, are often the source of representation. Significant aspects of the reasoning that scientists engage when developing their research can be accommodated in terms of the framework just introduced. Scientists explore the consequences from the models they use, and even when the information is incomplete, which it typically is, they try to reason about the possible scenarios in terms of the resources offered by the models at hand.

Partial Structures, Structural Objects, and Informational Structuralism

The partial structures framework offers a very natural setting for developing a different type of informational structuralism—both informational structural realism and informational structural empiricism. Crucial to structuralism is the emphasis on structure rather than on the nature of the objects under consideration. It does not matter which objects one considers—either because their nature is unknowable (as the epistemic structural realist insists) or because structure is all there is (according to the ontic structural realist). How can we make sense of the idea that objects do not matter?

One possibility is to insist that structural objects—objects on a structuralist reading—are only characterized in virtue of the relations they bear with other objects in a structure. Objects are what they are in virtue of these relations. As a result, structural objects lack intrinsic properties. As Lewis notes, "a thing has its intrinsic properties in virtue of the way that thing itself, and nothing else, is" (1983, 197). The situation is different for extrinsic properties, though, given that "a thing may well have these [extrinsic properties] in virtue of the way some larger whole is" (197). In the end, "the intrinsic properties of something depend only on that thing; whereas the extrinsic properties of something may depend, wholly or partly, on something else" (197). If structural objects only have extrinsic properties, it really does not matter which objects we consider—as long as the objects in question bear the appropriate relations with other objects in a structure, nothing more is needed.

Consider, for example, the number 3 in an arithmetical structure. As a structural object, that number is what it is in virtue of being the successor of number 2 and the predecessor of number 4. According to the structuralist, the nature of that number, beyond the context of the structure, is not relevant. In fact, it is not even clear that there is a fact of the matter as to what that number is outside the structure, given that whatever properties that number has are only specified in relation to other items in the structure (see Resnik 1997).

Using the partial structures framework, a structural object can be formally characterized via the concept of a partial equivalence relation. This is an equivalence relation—that is, a relation that is reflexive, symmetric, and transitive—but is defined for partial relations. So, if R is a partial equivalence relation, then the R_1- and R_2-components of R satisfy the three conditions of an equivalence relation, but it is left open whether the R_3 components also do. (Clearly, if they do, we obtain the full concept of an equivalence relation.) A partial equivalence relation determines a partial equivalence class, that is, a class of things for which it does not matter which member in the class one considers; any one of them will satisfy the conditions for a partial equivalence relation.

Structural objects are then objects in a partial equivalence relation; which relation it is depends on the particular context under consideration. For example, in nonrelativistic quantum mechanics, a relation of partial indistinguishability—that is, indistinguishability with respect to certain properties (defined by quantum mechanics)—allows us to express the point that it does not really matter which electron we consider; replacing one electron with another does not change the state the quantum system is in (French and Krause 2006). In this sense, electrons as individual objects do not play any role; what is significant are the relations electrons bear with other things, and in particular the fact that they are (partially) indistinguishable from other electrons.

Structural objects also bear relations with other objects, and this allows us to transfer information from one domain to another. In fact, the various kinds of partial morphisms discussed in the previous section illustrate information-preserving mappings among such objects. Once we establish that two partial structures are, for example, partially iso-morphic, properties that hold (in the sense of being quasi-true) for structural objects in one partial structure will then also hold for structural objects in the other structure, and vice versa. If two partial structures are partially homomorphic, then such transferring of information goes in one direction only. This provides a mechanism for transferring information among structural objects, without requiring that the nature of such objects be settled.

Now suppose the partial structures in question are interpreted in a realist way. That is, the structures are understood as offering a faithful description of the relevant physical system (including the unobservable properties of the system). In this case, we obtain a form of structural realism: a realist reading of the relevant partial structures in which the nature of the structural objects is left entirely open. Such openness can be interpreted in two ways: (a) If it is thought to be an epistemological matter, a limitation in our ability to know the nature of the relevant objects, we obtain a form of epistemic structural realism. (b) If the openness is thought to be an ontological matter—the objects in question have no underlying nature—we obtain a form of ontic structural realism.

In this way, there is room in this framework for capturing robust forms of structural realism.

But we need not interpret the partial structures under consideration in a realist way. Suppose the partial structures are understood as not offering a complete description of the relevant physical system, and that only the observable properties of the system are successfully characterized. With regard to the unobservable properties, given familiar underdetermination arguments, we are unable to settle what is really going on. In this case, we can remain agnostic about the unobservable features of the physical system, restricting our commitment to the observable aspects of the system. Given that the description of the system is made in structural terms—in fact, in terms of partial structures—and since only the observable parts are effectively captured, we have here a form of structural empiricism (Bueno 1999).

The framework suggested here thus allows us to express different formulations of structuralism in philosophy of science. Despite the significant differences between these views, they all emphasize the importance of information and information-preserving mechanisms in science. In fact, as noted above, one of the motivations for introducing partial structures was precisely to accommodate the crucial role played by the partiality of information in scientific practice, as well the various ways of transferring information across various domains of inquiry (via suitable partial morphisms). In this way, we have an alternative way of formulating informational structural realism. The new formulation allows us to obtain the distinctive features of epistemic and ontic structural realism, and, differently from Floridi's account, it makes both of them robust forms of realism. Moreover, a form of structural empiricism is also obtained. Here is not the place, of course, to argue for which of these versions of structuralism should be preferred. My point is to suggest a different framework in which Floridi's insightful account of informational structural realism can be better articulated.

Conclusion

Informational structural realism is a very significant proposal. Its emphasis on information and structure is exactly right. The issue of how to motivate a form of realism in this context is more delicate, though. What I have offered here is an alternative framework to the one proposed by Floridi—a framework that, in principle, can realize better than Floridi's own proposal the integration of structure and information in a realist setting. By invoking partial structures and partial morphisms, it is possible to make such combination and still preserve the distinctive features of epistemic and ontic structural realism. But one can also interpret the relevant structures in an empiricist way, thus obtaining a form of informational structural empiricism. In the end, even within the informational structuralist family there is a fair bit of pluralism.

Acknowledgments

My thanks to Steven French for extremely helpful discussions.

References

Bueno, Otávio. 1997. "Empirical Adequacy: A Partial Structures Approach." *Studies in History and Philosophy of Science* 28:585–610.
———. 1999. "What Is Structural Empiricism? Scientific Change in an Empiricist Setting." *Erkenntnis* 50:59–85.
———. 2000. "Empiricism, Mathematical Change and Scientific Change." *Studies in History and Philosophy of Science* 31:269–96.
Bueno, Otávio, and Edélcio de Souza. 1996. "The Concept of Quasi-Truth." *Logique et Analyse* 153–54:183–99.
Bueno, Otávio, Steven French, and James Ladyman. 2002. "On Representing the Relationship Between the Mathematical and the Empirical." *Philosophy of Science* 69:497–518.
da Costa, Newton, and Steven French. 2003. *Science and Partial Truth*. New York: Oxford University Press.
Floridi, Luciano. 2005. "Is Semantic Information Meaningful Data?" *Philosophy and Phenomenological Research* 70:351–70.
———. 2008. "A Defence of Informational Structural Realism." *Synthese* 161:219–53.
French, Steven, and Décio Krause. 2006. *Identity in Physics*. Oxford: Clarendon Press.
French, Steven, and James Ladyman. 1997. "Superconductivity and Structures: Revisiting the London Account." *Studies in History and Philosophy of Modern Physics* 28:363–93.
———. 1999. "Reinflating the Semantic Approach." *International Studies in the Philosophy of Science* 13:103–21.
Grosser, Morton. 1979. *The Discovery of Neptune*. New York: Dover.
Ladyman, James. 1998. "What Is Structural Realism?" *Studies in History and Philosophy of Science* 29:409–24.
Lewis, David. 1983. "Extrinsic Properties." *Philosophical Studies* 44:197–200.
Mikenberg, Irene, Newton da Costa, and Rolando Chuaqui. 1986. "Pragmatic Truth and Approximation to Truth." *Journal of Symbolic Logic* 51:201–21.
Resnik, Michael. 1997. *Mathematics as a Science of Patterns*. Oxford: Clarendon Press.

van Fraassen, Bas C. 1980. *The Scientific Image*. Oxford: Clarendon Press.
———. 1991. *Quantum Mechanics: An Empiricist View*. Oxford: Clarendon Press.
———. 2008. *Scientific Representation: Paradoxes of Perspective*. Oxford: Clarendon Press.

9

WHY INFORMATION ETHICS MUST BEGIN WITH VIRTUE ETHICS

RICHARD VOLKMAN

I can think of no better way to begin an essay on the information ethics (IE) of Floridi and Sanders than with the words of Emerson: "Every surmise and vaticination of the mind is entitled to a certain respect, and we learn to prefer imperfect theories, and sentences which contain glimpses of truth, to digested systems which have no one valuable suggestion" (Emerson 1982b, 77). While there are glimpses of truth in IE, its inherent foundationalism and extreme impartialism and universalism cannot do full justice to the rich data of ethical experience. To show this, it is requisite to unpack an alternative account that is explicitly antifoundationalist and fully agent-relative. I shall present a virtue ethics alternative to IE that conceives our ethical obligations as emerging from within our various particular perspectives instead of being imposed on us from the outside. Obviously, it is not possible to rigorously explore every nuance of these competing perspectives, but the point is a metaphilosophical comparison of these schools' defining commitments. In considering the alternatives side by side, it will become clear why the central contentions of IE, including especially the principle of ontological equality, must either express commitments grounded in the particular perspectives we already inhabit, or be without rational or ethical force *for us*.

Floridi proposes IE as a solution to an alleged "foundationalist crisis" in computer ethics, emerging from the realization that information itself is increasingly an object of concern in ethical discourse. Floridi argues that putting information at the heart of ethics results in an "ontocentric" theory, concerned with what things are, in contrast to the "situated action ethics" of the standard utilitarian, contractualist, and deontological alternatives. This ontocentrism issues in a very broad understanding of what commands moral respect: "From an IE perspective, the ethical discourse now comes to concern information as such, that is not just all persons, their cultivation, well-being and social interactions, not just animals, plants and their proper natural life, but also anything that exists, from paintings and books to stars and stones; anything that may or will exist, like future generations; and anything that was but is no more, like our ancestors" (Floridi 1999, 43). This perspective issues in the "ontological equality principle," which asserts that "any form of reality (any instance of information), simply for the fact of being what it is, enjoys an initial, overridable, equal right to exist and develop in a way which is appropriate to its nature" (44). Like virtue ethics, and in contrast to other alternatives, IE directs us to become good stewards of the "infosphere" by adopting an appropriate concern for what things are and what they are becoming, such that we construct a better world overall: "Ethics is not only a question of dealing morally well with a given world. It is also a question of constructing the world, improving its nature and shaping its development in the right way" (Floridi and Sanders 2005, 195).

In light of its emphasis on construction, the distance between virtue ethics and IE is not vast, but the fissure that separates them runs deep. Indeed, redirecting attention to good construction may be the most important contribution of the information turn, and the recovered emphasis on *poiesis* will be enthusiastically embraced by the proponent of virtue ethics. However, while virtue ethics can and should agree with Floridi and Sanders (2004) that specifying agency is relative to one's level of abstraction, such that humans constitute but a subset of those information entities that are agents, IE incorrectly supposes that there are judgments regarding the being and flourishing of information entities that are not bound to the perspective of some agent, and that these judgments can enter into human decisions about what to do and who to be. As we shall see, virtue ethics is not *narrowly* anthropocentric and rejects shallow subjectivism, but it does assert a fundamental role for agents in the evaluation and construction of the universe. While virtue ethics concedes that any information entity (not just humans) *might* merit respect and admiration for being/becoming what it is, that respect has to be earned by honest argument from information available within the perspective of the agent who admires and respects, rather than being perceived from the point of view of the universe, a point of view that is not the point of view of any actual agent and may in fact be antagonistic to the integrity of actual agents.

Of course, the refusal of virtue ethics to abstract away from the messy particulars of lives as they are actually lived makes it difficult to find one's bearings in a manifest plurality of considerations. In contrast, IE purports to solve a foundationalist crisis in computer ethics by getting at a single notion that can make sense of all the rest. We should be skeptical, however, of the notion that ethical discourse and understanding is served by the search for such foundations. "Theory typically uses the assumption that we probably have too many ethical ideas, some of which may well turn out to be mere prejudices. Our major problem now is actually that we have not too many but too few, and we need to cherish as many as we can" (Williams 1985, 117). It would be most ironic if a philosophy that asserts the fundamental value of information as such should be eager to abstract from the concrete information needed to make one's way in the world. We experience a world of incommensurable and conflicting values, and we must be careful not to smooth this over in first principles of ethical theory. Virtue ethics eschews foundations and proposes we start where we are. This approach will occasion objections of relativism. Just how diverse human virtues actually are is an empirical matter, but in principle virtue ethics embraces a narrow sort of relativism as against the extreme impartialism and universalism of IE. We shall see that such relativism does not run contrary to our actual ethical experience, and even accounts for data the impartialist alternative cannot.

History, Historicism, and Context

Floridi repeatedly intimates that IE is simply the next logical step in a long historical dialectic from the anthropocentric, particular, partial, and parochial toward universal and impartial verities in ethics. "Investigations have led researchers to move from more restricted to more inclusive, anthropocentric criteria and then further on towards biocentric criteria. As the most recent stage in this dialectical development, IE maintains that even biocentric analyses are still biased and too restricted in scope" (Floridi 2002, 297). However, the siren of historicist reasoning must be resisted here as elsewhere; as usual, a closer inspection reveals this is, as Popper puts it, "merely one of the many instances of metaphysical theories seemingly confirmed by facts—facts which, if examined more closely, turn out to be selected in the light of the very theories they are supposed to test" (1985, 300). To be sure, there is a certain tendency among students of philosophy to conceive the unfolding of the history of moral theory as progress from the narrow parochialism of virtue ethics, to the aspirations of universal and impartial theories like deontology and utilitarianism, to an even more universal and impartial understanding in environmental ethics. Reading history this way, IE looks like the next and obvious step.

But this is a very selective and distorted history. Throughout the Middle Ages it made sense to "take for granted, as we take the ground we stand on, the premise that the most important function of ethical theory is to tell you what sort of life is most desirable, or most worth living. That is, the point of ethics is that it is good for you, that it serves your self-interest" (Hunt 1999, 72). Even since the Enlightenment there has remained an important strand of thought challenging the notion that ethics requires impartiality and universality—to name just a very few obvious and illustrious cases, we have Kierkegaard and Nietzsche and Emerson in the nineteenth century, and Foot and Williams and Nussbaum and the whole resurgence of virtue ethics since. With this pedigree, unrepentant opponents of universalism and impartialism cannot be cast as mere fringe characters. Furthermore, this alternative historical emphasis indicates that the drive to be universal and impartial is a newcomer to ethical discourse, and it remains to be seen whether it will ultimately be judged a wrong turn.

Even the figures that seem to be most friendly to IE are at best mixed in their support. Plato and the Stoics are often cited as precursors to IE (e.g., Floridi 2006, 31), but both traditions appreciated the importance of grounding moral concern in the good life. The Stoics exerted tremendous influence on Nietzsche and Emerson, both of whom regard the rejection of impartialism and universalism as not only consistent with but even required by an appropriate reverence and affirmation with respect to the unfolding of Nature and *logos*. In light of Floridi's repeated emphasis on impartialism and universalism in his articulation and defense of IE, and especially the principle of ontological equality, any honest evaluation of IE needs to address itself to the ethical tradition that eschews impartialism and universalism while affirming the constructionist vocation of humanity. In particular, we need to ask what it can mean to be a good creator if one is unwilling or unable to discriminate between the information entities that merit respect and admiration and those that have not earned this status. If the act of creation is always an effort to inflict one's self on the world, as Nietzsche would have it, or if Nature is a manifestation of the self-reliant creation of one's life, as Emerson would have it, then not everything automatically commands respect as the thing it is. To evaluate IE, we need to critically interrogate the universalism and impartialism it takes to the extreme. "Unlike other non-standard ethics, IE is more impartial and universal—or one may say less ethically biased—because it brings to ultimate completion the process of enlargement of the concept of what may count as a centre of moral claims, which now includes every instance of information, no matter whether physically implemented or not" (Floridi 1999, 43). If impartialism and universalism turn out to be undesirable in themselves, at least when carried beyond their appropriate domains, then much of ethics since the Enlightenment has been a mistake, with IE as the most recent and most glaring example. In that case, the great historical contribution of IE will be to reveal once and for all the absurdity of extending impartiality and universality where they do not belong.

Impartialism and Universalism Within the Limits of Reason Alone

Of course, it would be absurd to suppose that a just account of ethics could entirely dispense with notions of impartiality and universality. After all, the referee acts wrongly who decides the game based on whom he likes better rather than the competitors' performance according to the rules of the game. While nothing could be more obvious, note that this reasoning is what Floridi calls "situated action ethics." It is no coincidence that notions of impartiality and universality make their greatest impact in the history of philosophy just as constructionist concerns fade in favor of moral theory that focuses on action. While rules of conduct and policy are certainly bound in their formulation and application by considerations of impartiality and universality, policy and rules of conduct are but the smallest part of normative discourse as a whole. Let us mark off this part of normative discourse by the designation "politics," since it deals with matters of policy rather than the constructionist concerns of virtue and flourishing. This distinction reveals that impartiality and universalism are appropriate in circumstances of conflict or competition, especially involving relations of strangers or neighbors in contrast to friends and family. Impartiality and universalism are ideals associated with justice—the virtue concerned with giving others their due. But justice only arises as a concern when one might be tempted to give less than others are due. "When men are friends, there is no need of justice" (Aristotle 2009, VIII:1). Not only is justice unnecessary among friends, it is positively unwelcome, as are the impartialism and universalism that characterize it.

It has become a commonplace, starting especially with Kant and Bentham, to suppose that an attitude of universal impartiality is simply a requirement of reason itself. On this view, it is irrational to cast one's self as an exception to some universal rule or policy without some justification, since that would involve asserting an arbitrary difference. If it is wrong for you to lie to me, then I have to admit it is wrong for me to lie to you; if my pleasure counts in what should be done, then I have to admit that your pleasure counts too. If there is some reason why I count and you don't (or vice versa), then I need to be prepared to show the difference that justifies the difference in treatment. If ethics is conceived as casting about for just such rules of conduct, on the grounds that reasons for action must apply equally for everyone as reasons, then it follows that ethical reasons must be of the sort one is willing to apply equally to everyone, without special reference to one's own concerns or circumstance. In this way, impartiality and universality of reasons for action interpenetrate and reinforce one another as the conditions of rational principles for guiding conduct.

But this focus on reasons for action, as if the agent is a calculator or database of rules, misses most of what really matters in ethics, and it is not at all the only way to conceive rationality. Nussbaum unpacks the virtue ethics alternative in her essay "The Discernment of Perception: An Aristotelian Conception of Rationality." Since virtue ethics is especially concerned with good construction of one's self in one's actual circumstance, rather than narrowly worrying about activity in the abstract, the creative act of choice must embrace all the particulars that make one's life make sense as the particular unfolding story it is. As Nussbaum argues, "excellent choice cannot be captured in general rules, because it is a matter of fitting one's choice to the complex requirements of a concrete situation" (1990, 71). It is significant that proponents of situated action ethics tend to illustrate ethical choice by means of a few thin descriptions of some agent (often named A or B) in some abstractly specified situation; in contrast, Nussbaum indicates that it takes a novel to properly capture ethical choice as it actually takes place. To boil down all the facts into the thin descriptions of casuistry is to have already done all the real ethical work behind the scenes, instead of confronting the messy multitude of competing claims and appeals we find in every moment of real life, and Floridi is surely correct that the transformative circumstance of the information age speaks against treating all the ethical work regarding who I am and who I aspire to become as already complete once and for all behind the scenes. Construction matters if we are to understand ethics in the Information Age.

Having identified construction as the vocation of ethics in the Information Age, we need to sort out what is to be constructed and along what dimensions. Aristotelian virtue ethics rejects the notion that everything that is good can be boiled down and weighed along a single dimension; the only way to face the radical pluralism and incommensurability of goods as they appear in our ethical experience is to learn how to process all this information in the complicated way indicated by perceptual awareness. That is, one has to learn how to *see* good and bad: "The discernment rests with perception" (Aristotle, qtd. Nussbaum 1990, 66). Deliberation informed by right perception is something one does, not a rule one follows, and it is only learned by disciplined practice under the tutelage of good mentors and exemplars; it cannot be summed up in a catalogue of rules any more than one can become a great musician by reading a book. "Good deliberation," says Nussbaum, "is like theatrical or musical improvisation, where what counts is flexibility, responsiveness, and openness to the external; to rely on an algorithm here is not only insufficient, it is a sign of immaturity and weakness" (1990, 74).

The conception of rationality assumed by situated action ethics ignores much of concrete concern in the unfolding story of actual life, and such an ethics is too impoverished to do justice to our ethical experience of values of good construction. If we are concerned fundamentally with making something of ourselves, then we cannot accept action-oriented ethics. This is a point most famously made by Bernard Williams in his analysis of integrity (Williams 1988). Williams offers cases in which an agent is confronted with an action that goes against some ethical commitment that has come to constitute a central part of his identity, but which would surely result in a better outcome in terms of some favored conception of the "overall good" advocated by consequentialism. The point is that the agent's integrity is not accounted for in the consequentialist tallying of the "overall good," no matter what one's specific value theory says, since the consequentialist approach to ethics implies the agent who is committed to doing the right thing must be prepared to abandon his own constitutive commitments whenever an objective tallying of the overall good requires it, and being so prepared is already to have abandoned one's integrity. One's integrity cannot be simply weighed against other considerations as if it was something commensurable with them. Being prepared to do that is already to say one will be whatever the utilitarian standard says one must be, which is to have already abandoned one's integrity. The priority of the particular in Aristotelian ethics emphasizes that our moral experience is shot through with this sort of radical incommensurability, and constructing a good life out of such material requires real sensitivity and judgment. Rules and policies are no substitute for this.

In my experience, proponents of situated action ethics are impatient with this line of reasoning. This is generally because they misunderstand the nature of the argument; they think it attempts to assert some contradictory observation against deontology or utilitarianism, as if the point was that utilitarianism is logically incapable of giving the right answer in Williams's cases. (Or worse, they mistake Williams's point to be that integrity trumps every other consideration; anyone who thinks that needs to go back and reread the previous paragraph more carefully.) So, they make great efforts to show how their favored theory manages to give the right answer after all, and we get tortured explanations showing that utilitarians can say it is wrong to shoot one Indian to save twenty others, or whatever. But this utterly misses the point. It is exactly that sort of misplaced puzzle-solving rationality that is the problem. Williams makes it clear that his cases are supposed to be real moral quandaries. No one can say with confidence whether or not it would be right to shoot one Indian to save twenty. Respecting one's integrity requires that one muddle through and *see* an impossible circumstance for the impossible case it is. That does *not* mean integrity overrides other concerns; it means that the relevant concerns cannot all be weighed alongside each other on a common scale. What's more, as Nussbaum emphasizes, it is not at all obvious that the right answer has to be the same no matter who the particular agent is, no matter who the particular Indians are, and no matter how these are related. What is the actual history that puts one in this circumstance of choice? What does it mean for the characters of the story to be shot? What does it mean for the shooter? The significance of these matters cannot be settled by the guidance of any simple algorithm without violence to one's integrity as a deliberative agent, which means that one's integrity is not accounted for by any such algorithm. Therefore, assuming, as most would grant, that integrity has any value at all, whatever else the algorithm might be good for, it does not account for everything of ethical value. No such algorithm captures the whole of ethics. However things turn out for the proponent of situated action ethics, these arguments are particularly telling against IE's uncritical acceptance of impartialism and universalism, since IE purports to value construction, and integrity is a constitutive virtue of a valuable structure.

To understand the ethical cost of abandoning the concrete perspective of actual agents, it helps to reflect on the sentiment expressed by Ivan in *The Brothers Karamazov* regarding the proposal that an eternal harmony of the universe answers the problem of evil. The usual conclusion of the Argument from Evil is that God does not exist, since an all-knowing, all-powerful, all-good God would not permit the evil we see in the world. One standard retort, captured powerfully and poetically in the answer from the whirlwind of the book of Job, is that God knows what He is doing even if the particulars of God's plan are often beyond our limited understandings. The usual countermove says assertions of God's plan are question begging, but Ivan makes a much more subtle and challenging response. He does not mean to deny God's existence; he means to indict God for crimes against humanity, and especially against innocent children. Given everything he knows and believes and sees, it is simply not acceptable to trade off the suffering of innocent children against even an eternity of perfect harmony and perfection. In asserting this, Ivan concedes that the trade-off might make sense to God. It may even make sense to certain humans. But Ivan wants none of that. "Perhaps it really may happen that if I live to that moment, or rise again to see it, I, too, perhaps, may cry aloud with the rest, looking at the mother embracing the child's torturer, 'Thou art just, O Lord!' but I don't want to cry aloud then. While there is still time, I hasten to protect myself, and so I renounce the higher harmony altogether. It's not worth the tears of that one tortured child" (Dostoevsky 2010, chap. 35).

Whatever may be true from some other perspective, Ivan cannot countenance the calculation that permits children to be thrown to dogs. To come to appreciate the calculation from God's point of view, even if possible, would not settle the matter in God's favor, and it would in fact constitute an ethically disastrous failure of integrity, as it would mean Ivan's ceasing to be what he is and becoming what he is not, what he in fact abhors. Even if Ivan's perspective and his perception are in some sense more limited and parochial than the all-knowing God, those limits are themselves constitutive of Ivan and of what it means to be successful as being what he is. As so often happens, the point is captured by Aristotle in an observation that at once appears perfectly banal and utterly profound: "No one chooses to possess the whole world if he has first to become someone else" (Aristotle 2009, IX:4). An ethics that adopts some point of view other than the agent's own may not be relevant to the agent's evaluation of things. However things may seem from the perspective of the principle of ontological equality, there is no reason to assume it has anything to do with *me*.

Although I cannot succeed in my life by becoming someone else, it is equally true that my own success depends on extending my self by including others in my very constitution. By far the most celebrated and widely discussed counterexamples to the pervasiveness of impartiality and universality focus on friendship and similar close personal relationships, and with good reason. In cases of love and friendship, "the agent's own historical singularity (and/or the historical singularity of the relationship itself) enter into moral deliberation in a way that could not even in principle give rise to a universal principle, since what is ethically important (among other things) is to treat the friend as a unique nonreplaceable being, a being not like anyone else in the world" (Nussbaum 1990, 72). We have already noted that friendship excludes justice, and we are now in a position to locate the crucial difference in the very being of the friendship. While anything like a complete account is well beyond the scope of this chapter, even a sidelong glance in the direction of friendship reveals not only what mere politics misses but also what virtue ethics has to contribute. Specifically, we can begin to sketch the ontological orientation of virtue ethics in unpacking friendship, revealing that partiality and particularity do not entail a narrow conception of self and other or a shallow account of the relations that bind us together.

On Aristotle's view, the friend is a "second self," in a quite literal sense that can be illustrated by a metaphor (inspired by Emerson 1982a): What I am and how I relate to others can be imagined as the expanding ripples on the surface of a pond upon the introduction of a stone. The boundaries that define an individual are the circles pushing out across the surface (representing the whole of the infosphere, if you like), with other individuals similarly represented as circles pushing out from their own distinct points of entry. The whole of one's life is the history of those circles as they expand through the pond, including especially the complicated emergent properties that issue from encounters with the ripples that are other lives. The resulting patterns become hopelessly intertwined and fused, and it becomes impossible to distinguish any precise boundaries. Whether a given part of the pond "belongs" to one or another wave becomes a hopeless question, but it remains that each circle expands from its own center, and that it pushes out to make more sophisticated patterns that are part of the story of expansion from that center. To tell the story of these waves, it is more than a bit helpful to identify the entrance of the stone as the "center of narrative gravity," to use Dennett's (1992) apt expression, which is all we need the self to be in a virtue ethics account. (Of course, the self might be something more substantial, but I take it to be *at least* a center of narrative gravity.) As the metaphor illustrates, *selves overlap*, and this is a matter of fundamental importance for the analysis of those human relations that are not matters of politics.

To understand one's life is to make sense of its constitutive elements and their relations in coherent whole, and that includes the people whose lives are literally part of one's self. The friend is a second self, since there is no distinguishing one's own interests from those of one's friends. There is no question of whether doing something valuable for one's son is a benefit for him or for one's self, any more than it makes sense to ask whether a tax break for a married couple filing jointly is a legal benefit for the husband or for the wife. Thus, the friend is valued for her own sake, and not as an instrument or from duty. Make no mistake about it, there is excellent reason to make friends; friendship is a construction that makes each of us greater than we could be apart. It follows that I have excellent reasons to make myself the sort of thing that can enter into these relations, and this is a sufficient consideration to underwrite all the various social virtues (e.g., honesty, compassion, benevolence) without appeal to impartiality or universality.

These points about the relation of one's self to one's friends apply equally well to the manifold other goods that constitute the good life. The contours of one's life story are composed of the various projects, values, and commitments that interact to define the narrative center of gravity, like so many stones in the pond. These considerations cannot be reduced or boiled down without distortion or extirpation of real pieces of one's self. "If thy eye offend thee, pluck is out" is one imperative; "Learn to see well" is quite another. Only the latter is consistent with the full flourishing of one's self as having all good things in the right proportion, which is the central concern of *eudaimonia*. Safety is good, but it is not the only good. Hence, I must become the sort of thing that cares not too much for safety and not too little for safety—I must be courageous. Pleasure is good but, if we agree that safety is also good and not reducible to pleasure, pleasure is plainly not the only good. Hence, I must become the sort of thing that cares not too much or too little for pleasure—I must be temperate rather than gluttonous or puritanical. Even anger is a good, in the right proportion, defining good temper against slavish submissiveness. But too much anger, or inappropriate anger, or anger that lingers makes one a hothead or short tempered or sulky, and these traits threaten to upset one's navigation through the crashing waves in the great sea of competing claims of value. This is the level of description adopted by virtue ethics, and these *observations* may be obscured if we reflect on the stuff of our lives from any other level of description.

The imperative to make something of my self is not chosen; it is thrust upon me by being what I am. It is a mistake of the highest order to suppose that this project of construction can be simplified by theory, or that there is an imperative standing outside the constructionist imperative that conditions it without entering into it. Justice is *a* good. Benevolence is *a* good. These are virtues that can be affirmed from within the viewpoint described above. But the imperialism of these considerations that opposes one's ability to become who one is must be opposed by the whole project. My life is already overfull in crafting for myself a character that manifests all the virtues (*including* justice and benevolence) without imagining a requirement that imposes further obligations of some alleged outside authority. We should say of impartiality and universalism in ethics what Foot suggests regarding the related notion of "the best outcome" as it relates to benevolent concern for others' welfare: "It is not that in the guise of 'the best outcome' [the idea of maximum welfare] stands *outside* morality as its foundation and arbiter, but rather that it appears *within* morality as the end of one of the virtues" (1988, 238; emphasis in original). To suppose otherwise is to suppose we have found the universal acid that permits us to dissolve all good things into a single thin solution for weighing and comparison. No, pleasure is not such a universal acid. Nor is information.

Obviously, the observation that neither pleasure nor information can serve this role is a substantive premise that could be disputed, but this deep pluralism about value is a central feature of the virtue ethics perspective, and IE generally accepts value pluralism even as it strives to maintain its extreme impartiality and universality. While calling attention to this tension in

IE, most of this description of virtue ethics is perfectly congenial to the information turn. Nothing in virtue ethics supposes that the agents are anything but information entities, or that only human agents are agents. The representation of agents as interacting waves in a field of possible interactions is perfectly suited to elaboration in terms of information. However, the analysis so far indicates the sense in which IE's abstract theorizing and searching for ethical foundations may expresses a disdain for our information-rich lives. The level of abstraction at which ontological equality holds is not the perspective of our actual lives, and it is not obvious that it generates any reasons for us at the perspective relevant to ethics. Virtue ethics proposes that, if we take the information turn, we do best to start where we are instead of positing external and information-impoverished points of view as foundational. "The foundationalist enterprise . . . has now generally been displaced in favor of a holistic type of model, in which some beliefs can be questioned, justified, or adjusted while other are kept constant, but there is no process by which they can all be questioned at once or all justified in terms of (almost) nothing" (Williams 1985, 113). Let us forget foundations in ethics, and instead start our construction project where we are. If we start where we are, we discover ourselves to be powerful information-processing systems whose greater perfection consists in ordering and understanding and appreciating and creating from the materials at hand.

You Can't See Nothing from Nowhere

Ontological equality supposes that all information entities merit respect and enjoy a (minimal, overridable) right to flourish as the things they are. But this can only seem to be the case if we imagine a perspective that is not tied to any particular history or situation—what Sidgwick famously describes as "the point of view of the universe." In contrast, if we start where we are, as the particular characters we are becoming, we can be prepared to affirm the value of any information entity we encounter, and it may turn out that every information entity merits my respect and should flourish as the thing it is, in which case the object-level judgments of virtue ethics and IE may have an identical extension. But that is certainly not the way things appear from the start. As soon as one appeals to a sufficiently concrete level of abstraction, we find a raucous competition between the flourishings of various info entities, and even a competition between alternative descriptions of the same info entity.

Before proceeding, we need to acknowledge the role in IE of particular applied ontologies in the formulation of policies and life plans. Ontological equality is not proposed as a rich guide to action, and the point of view of the universe was never intended to supplant our particular points of view. Rather, ontological equality was intended as supplement and corrective to the tendency of particular viewpoints to be totalizing and tyrannical. "The description of the specific essence of classes of information entities is a task to be left to a plurality of ontologies . . . IE relies on the agent's knowledge for the implementation of the right action" (Floridi 1999, 45). It may seem that there is no deep conflict between virtue ethics and IE after all, since the ontologies that inform our actual conduct are presumably the constitutive considerations emerging from the personal point of view described above. "I am using 'ontology' to cover the outcome of a variety of processes that allow an agent to appropriate (be successfully embedded in), semanticize (give meaning to and make sense of) and conceptualize (order, understand and explain) her environment" (Floridi 2007, 5).

These various ontologies are related to the levels of abstraction one adopts in coming at a problem, since, "a LoA [level of abstraction] rather expresses (often only implicitly) a specific ontological commitment, which the agent may or may not endorse" (Floridi 2008, 191). The picture of IE that emerges from this is that various levels of abstraction supplement and sometimes guide and correct other levels of abstraction involved in the unfolding of deliberation as instruments for pursuing our various ends. Since it appropriates to itself everything from the various levels of abstraction in addition to embracing insights available from levels of abstraction that merely consider Being as such, IE is supposed to be more information rich and sensitive than the virtue ethics alternative, which is always mired in the personal points of view of the various burgeoning individuals. When we are pressed with concrete cases, however, it becomes apparent that the most abstract level of description either adds no information whatsoever to our deliberations or contributes outright misinformation.

The problem of preserving meanings across levels of abstraction is indicated in Nagel's observation, "I don't know what it would mean to wonder whether, *sub specie aeternitatis*, wanting a chicken salad sandwich was perhaps really preferable as a ground for action to wanting a salami sandwich" (Nagel 1986, 131). Clearly, what matters from one point of view may not matter from another. "It matters a great deal what we ourselves think about our selfhood and our possibilities; what a being who stands apart from our experiences and ways of life thinks seems to matter little, if at all" (Nussbaum 1995, 121). Floridi has sometimes resisted describing IE in such terms, but it is not clear what else to make of claims that IE requires "a disinterested judgment of the moral situation from an objective perspective" (Floridi 2006, 26). This language of perspective or point of view indicates that a shift from our antecedent standpoint is required in order to see the matter correctly, but it must not be assumed that what one sees from one vantage point can simply be added or extended to other ways of seeing the world. When we abstract away from particular contexts, we should no more suppose we still have a clear idea of what we mean than when we try to sort out the "real" relative merits of chicken salad and salami.

These concerns can be illustrated using Floridi's own examples. He describes a case in which a boy passes time by smashing the windshields of junk cars (Floridi 1999). The case stipulates that these windscreens will not be used or even seen by anyone in the future, so there is no question about the boy's actions in relation to utility or the interests of others. According to IE, the boy's actions are rightly condemned as vandalism, and Floridi asserts that this matches our intuitive sense of the case. Since the observation that the boy's vandalism is wrong makes sense only on the assumption of the intrinsic right of the windshields to be what they are, that is, only if we assume the principle of ontological equality, Floridi concludes this case is evidence of the value of IE. However, there is much to dispute in this case. First of all, it is hardly obvious that this is a case of wrongful vandalism. The proponent of situated-action ethics will surely claim that our observation is tainted by imagined uses of these windscreens or the property rights of whoever owns the junk cars; the virtue ethicist will complain that the story is far too thin to make any adequate judgment. Is the boy really just getting mindless kicks, or is he rehearsing his shot? How much time are we talking about? What are the alternatives open to him? What brought him here, and where is he going? There are myriad coherent stories in which it would be perfectly O.K. to smash things.

But the deeper problems with the case are matters of ontology. The description of "vandalism" presupposes that the windscreens are properly described according to some essence or nature that is violated in the act of smashing them. According to ontological equality, their integrity as windshields demands that we not smash them without reason. But it is a profound error to suppose that these things "really are" windshields from the point of view of the universe. The boy treats them as targets. In fact, the story posits that no one will ever treat them as windshields ever again. In that case, why should that description be privileged? Perhaps Floridi means not to privilege that description but merely to treat it as one ontology alongside others. But in that case, there can be no way to respect this entity for what it is and promote its flourishing, since its flourishing under one ontology excludes its flourishing under another. From the point of view that commands us to offer such respect, the thing has no nature and no flourishing. From a point of view such that it has a nature and can flourish as the thing it is, it does not command any particular respect (by hypothesis, it will never fulfill its function as a windscreen), and it may even cry out to be smashed (the flourishing of a target). Perhaps the boy behaves wrongly unless he smashes it!

This may sound like an objection Floridi has repeatedly tried to address—that IE either is not a precise or clear guide to action or leads to relativism. However, my point is much deeper: Whatever observations may seem to make sense from the point of view of the universe are irrelevant to the project of a human life. IE can guide action, but it guides us badly by suggesting an equality that does not exist at the *relevant* level of abstraction. While previous critics of IE have generally admitted IE's impartialism or universality as important values, virtue ethics disagrees on these first premises of the IE project, and that transforms the nature of the critique. For example, Stahl questions IE's claims to universality by noting that one's level of abstraction expresses an ontology and, "as the choice of the LoA [level of abstraction] is not determined and given that it determines ontological commitments, it seems that an agent is free to choose, albeit indirectly, the ontology of a phenomenon. This allows the interpretation that an agent gets to determine which information entities exist (are relevant) and therefore need to be considered in ethical decision making" (2008, 100). On this point, Floridi responds that, although there is no single level of abstraction that is appropriate for all purposes, some levels of abstraction are more or less appropriate for given purposes. Thus, he maintains against Stahl that worries about relativism are misplaced because one's level of abstraction may be more or less appropriate to a task at hand, and the level of abstraction that reveals ontological equality happens to be useful for our ethical purposes. "So, the position held by IE is that, when it comes to talk about ethical issues in an ontocentric and more inclusive, non-anthropocentric way, an informational LoA [level of abstraction] does a good job" (Floridi 2008, 190).

But this will not do as a response to the concerns raised here. If we are willing to embrace a narrow relativism on this point, then there cannot be an ontocentric way to approach ethical issues that abstracts from the grounds of our discriminating things into the things they are, since what a thing is depends on our actual purposes and values (e.g., is it a target or a windshield), and it is exactly those purposes and values that determine which levels of abstraction are appropriate for any given deliberation. An informational LoA does not permit the discriminations that would ground conduct, since it transcends the actual concerns of situated agents, and it is not clear how anything gleaned from that perspective could provide reasons that must be taken seriously in one's situated deliberations, since the informational LoA is manifestly *not* appropriate for solving the problems seen from one's actual perspective. That is, if the appropriateness of a level of abstraction is given by one's purposes, then the informational LoA is not appropriate, unless one's purposes are exhausted in a concern for ethics in an "ontocentric and more inclusive, non-anthropocentric way." The fact that everything seems to be valuable when I ignore how things differently impact my actual projects and constitutive commitments can only ground a reason to treat all things as having intrinsic value as the things they are if I already have a reason to ignore those distinctions that make them appear as the things they are *for me*. But indiscriminately ignoring those distinctions would alter my constitutive commitments, which would change who I am, which means IE is not merely supplementing my perspective but really does ask me to abandon my perspective in favor of another.

The problem is not mere underdetermination or puzzlement about how to apply IE. In a range of concrete cases, the principle of ontological equality can *only* lead to misunderstanding one's circumstance. It cannot inform good deliberation. If I am a musician working on a song that includes a banjo track, should I add a bit of distortion to it? Although it is quite literally adding noise to the signal, knowing when to use distortion effects in music is plainly a matter of good construction, and I submit this knowledge is not advanced one bit by considering the effects of distortion on the banjo signal as an object of moral concern. To the contrary, this can only muddle a case that must be handled with a subtle sensitivity to the concrete facts of the case: What is the genre? What is the musical motive of the piece? Who is the audience? How does it all fit within the wider context of my musical tradition? What would distortion on the banjo track *mean* in this context? A brilliant move in bluegrass can be a mistake in jazz. While there is no formula or simple algorithm for all this, we must emphasize that this analysis does not entail a shallow relativism or subjectivism, and it is not an empty situation ethics. It is wrong to suppose that anything goes in a bluegrass jam. Improvisation is not about rules or policies, but there are mistakes in jazz. Or, alternately, if something is a mistake, then it is not jazz. Either way, there is something to perform well as the thing one is, but it involves the instant processing of vast amounts of varied information along numerous incommensurable dimensions. We call that sort of information processing *perception*, and it cannot be advanced by learning about the principle of ontological equality.

Notice how all this happily embraces conception in terms of information, but only if we appreciate the concrete narrative that confers meaning on the information entity in question. The banjo signal has no right in itself to be a banjo signal; that right must be *earned* in virtue of the role it plays in some coherent story. To suppose otherwise obstructs the story-telling that makes meaning. For some things, the contour of the story is forced upon us; for others, it is impossible to tell their story without anthropomorphizing and personification. My dog's chew toy sits on the floor. Is it flourishing as the thing it is? Or would it be better if my dog was chewing on it? Does the chew toy flourish as the thing it is by maintaining the integrity of its bone shape into the future, or by being torn to pieces by a dog? In contrast, suppose the *Mona Lisa* is on my floor. Now, there is no question that I should not permit my dog to chew on it, and there is no question that a sensitive agent will perceive the story of this information entity and what it means to respect its integrity and flourishing without anthropomorphizing or personification. Since there is no narrative center of gravity for the chew toy, it does not merit my respect, and cataloguing its trajectory through time and space is not meaningful. A chronicle is not a history, because it is not a story. But the story of the *Mona Lisa* centers on itself as a thing in its own right (and not just as the creation of Leonardo). It commands my respect and admiration for this reason, rather than as the birthright of its being as an information object. The importance of coherent storytelling is not meant to deny the reality of information in the absence of the storyteller; it points us to the fact that such information cannot be seen to matter. The information turn pushes us to acknowledge that not all information is equal.

Storytelling grounds the meaning and significance of information objects while maintaining priority of the particular. Brey (2008) argues that Floridi's defense of the intrinsic value of all information objects commits what is sometimes called the Birthday Fallacy: Everyone has a birthday; therefore, there must be a single day on which every person was born. As Brey tells it, Floridi (2002) supposes that there must one thing that everything with intrinsic value shares, that there must be a "minimalist and homogenous account" of intrinsic value. Floridi responds that IE attempts to "show in practice a way of conceptualizing Being informationally in such a way as to build a minimalist and homogenous account of all entities" (Floridi 2008, 193). So, the informational LoA reveals a commonality among all information objects that can explain their intrinsic value, and this hypothesis is confirmed by its actual usefulness in theory and practice. However, if a thing's being what it is and having value is relative to the competing stories we can tell about it, then the informational LoA will have left out what makes the thing what it is with whatever value it has. At a minimum, this represents an alternative to the IE explanation of value. Moreover, it indicates how the sources of value may be multifarious and incommensurable; just as stories bear a family resemblance that makes them all stories in the absence of a shared essence, so the values of things will be heterogeneous.

The objection to this approach seems to be that it is unabashedly anthropomorphic and relativistic. If one has already accepted that all of ethics needs to be utterly impartial and universal, that will probably seem like a serious objection. However, virtue ethics does not concede that this narrow relativism (which must not be confused with shallow, anything-goes subjectivism) is an objection. Rather, our longing for universality and impartiality belongs in politics, not ethics, and respect for the deep pluralism that implies relativism is a condition of constructing the good society, not an obstruction of it.

Sociopoiesis: Justice Means Competition Is Cooperation

Some information entities craft their own stories, including the stories that define them. These entities will always merit the respect described above, since they will have some center of narrative gravity that is at the same time a narrator and potential interlocutor. As already noted, there are no sharp boundaries between individuals so defined. Our lives quite literally overlap as expanding circles of influence intermingling with others. Sometimes, that intermingling is best conceived as forming an altogether distinct and coherent information entity with a story of its own—a friendship. In other cases, the story is more perspicuous that treats these waves as confronting and competing against one another. It is in these cases that politics and the virtue of justice enjoy their full significance. Most human interactions are neither entirely matters of friendship nor entirely matters of justice. It requires discernment to know which claims demand attention when.

Many virtue ethicists have incorrectly supposed that politics involves nothing more than "scaling up" the virtues as applying to society as a whole, and this seems to be Floridi's own conception of the virtue ethics account of the making of society—"sociopoiesis" (Floridi and Sanders 2005). However, this communitarian approach to politics mistakes society for friendship. Insofar as we really are sorting out relations that are not matters of friendship, good construction will not be the making of a single unity. Whose perspective, whose conception of the good, whose story should prevail? There is no outside perspective to answer this, and the resulting competition to impose one's conception of the good rather than being imposed upon is a Hobbesian war of all against all. If we really are in something like a competition, and if each of us really does merit respect, then politics is better understood in terms of a nonaggression pact rather than a universal republic. "The most schematic code against interference and mutual destruction may be enough for parties who merely have a shared requirement to live, not a requirement to share a life" (Williams 1985, 103). As Rawls (2005) has emphasized, liberalism finds its appeal in resolving the fact of "permanent reasonable pluralism." An extended society is not a friendship, club, community, or association. While there are indeed no sharp boundaries between persons as such, there are plainly distinct centers of narrative gravity that have stronger and weaker claims to various regions of moral space. Most of us most of the time recognize the legitimate sovereignty of others over what is their own, but conflict emerges at the unavoidably vague boundaries. Since even persons of perfectly good will are bad judges in their own cases, politics and government is the necessary business of drawing up and enforcing sharp boundaries that define our legitimate expectations with respect to what counts as my life, liberty, and property and what counts as yours. (This will be recognized as a sketch of Locke's *Second Treatise on Government*.)

Liberal institutions transcend mere nonaggression by establishing the only mechanism capable of leveraging the vast amounts of tacit, local, and distributed information that is needed for wise sociopoiesis. Liberal rights facilitate market relations that supplement friendships and supplant competition as combat, and market institutions harness the wisdom of crowds as information-processing technologies of the highest order. "Market institutions" must not be understood in a narrowly economic sense, since they include the distributed processing that constitutes the marketplace of ideas and underwrites progress as the condition of cultural and social evolution. This is the only way to proceed if sociopoiesis matters in the information age of complex and open societies. "It is in the utilization, in the mutually adjusted efforts of different people, of more knowledge than anyone possesses or than it is possible intellectually to synthesize, that achievements emerge that are greater than any one man's mind can foresee" (Hayek 1958, 193). A political order characterized by rule of law and liberal rights embraces impartiality and universalism in their appropriate domain, while establishing the conditions for the emergence of the good society. Within this framework, relativism, which amounts to authority in its original sense with respect to one's own life's story, will not offend against any obvious moral rights or duties.

Reverence from the Inside Out

Given their willingness to embrace local ontologies in evaluating the meaning and significance of states of affairs and entities, perhaps Floridi and Sanders can welcome the notion that individuals working from within their individual perspectives are best situated to find their way through their individual lives, while IE remains true at a level of abstraction suited to philosophy and as an external confirmation of certain of the insights informed by our personal points of view. However, they may balk at the notion that an understanding of our universe that proceeds from the inside out can do justice to our objective relation to the world. Perhaps they will maintain that there is something just wrong with virtue ethics insofar as it remains so unapologetically anthropocentric and egocentric. Against this, I submit that everything we can appreciate from the perspective of IE can be equally well appreciated without jumping outside our own skins. Virtue ethics is neither narrowly anthropocentric nor selfishly egocentric.

As Williams notes, "The word 'speciesism' has been used for an attitude some regard as our ultimate prejudice, that in favor of humanity. It is more revealingly called 'humanism,' and it is not a prejudice. To see the world from a human point of view is not an absurd thing for human beings to do" (1985, 118). It must be emphasized, however, that this does not entail a failure to appreciate and respect the myriad claims to value we encounter as our lives unfold in the concrete. "Our approach to these issues cannot and should not be narrowly anthropocentric" (Williams 1994, 47). If we remain open to whatever possibilities and arguments we may confront when we start where we are, there is nothing *narrowly* anthropocentric about starting where we are. Similarly, while every ethical concern has to be addressed to us as a concern *of us*, it does not follow that every concern is *reducible* to self-concern.

This responds directly to Floridi's defense of ontological equality, which "consists in shifting the burden of proof by asking, from a patient-oriented perspective, whether there is anything in the universe that is intrinsically and positively worthless ethically and hence rightly disrespectable in this particular sense, i.e. insofar as its intrinsic value is concerned" (Floridi 2008, 192). If foundationalism is wrongheaded and we are justified in starting where we are, then anthropocentrism is not mere conservatism or prejudice. Nothing is arbitrarily cast aside as worthless, for we do not assume anything that does not appear immediately in the first person, and everything that appears is subject to revision. We expect the value of things to be revealed in the light of a vigorous and open-minded scrutiny of our present concerns, since nothing that fails to speak to these concerns can ground the reasons we actually have as the things we actually are. So, the challenge should be clear enough: For any entity you like, go ahead and set out to show that *this* entity commands my respect by appealing to the particular projects and constitutive commitments that define my character in the story of my life. I am in no way denying that you may be able to do that for any entity you choose. What I am denying is that any one account can do this all at once for everything and everyone. Reverence has to be worked up from the inside out.

It is significant that Emerson and the American Transcendentalists stand at once as early advocates for reverence as due unto Nature and as the great champions of the heroic individualism that evaluates the world from the inside out. These are not mutually exclusive projects. Mere things are the appropriate occasion to reflect on what is highest, and that will entail the affirmation of life itself. "The stars awaken a certain reverence, because though always present, they are inaccessible" (Emerson 1982b, 37). For the poet or the philosopher, or better yet, for the heroic individual who aspires to be both, contemplation of Nature is the very stuff of religious experience. "We can never see Christianity from the catechism:—from the pastures, from a boat in the pond, from amidst the songs of wood-birds we possibly may" (Emerson 1982a, 232–33). It is not surprising that this attitude seems at odds with conceiving one's self as an individual apart from Nature. Indeed, standing before Nature tends to the dissolution of the individual consciousness, but this is where the real greatness of what it means to be a human individual is revealed: "Standing on the bare ground,—my head bathed by the blithe air and uplifted into infinite space,—all mean egotism vanishes. I become a transparent eyeball; I am nothing; I see all; the currents of the Universal Being circulate through me; I am part or parcel of God" (Emerson 1982b, 39). In this light, the integrity and self-reliance of the individual are expressions of reverence, for "God will not have His work made manifest by cowards" (Emerson 1982c, 176). Far from disproving one's individual greatness, our encounters with Nature underline and confirm the heroism implicit in becoming who we are. "The hero always bears the wilderness and the sacred, inviolable borderline within him wherever he may go" (Nietzsche, qtd. Thiele 1990, 21).

Seen in this light, it becomes apparent that Floridi and Sanders are wrong to accuse the ethical individualist thread of virtue ethics of "moral escapism" (Floridi and Sanders 2005, 198). To the contrary, as Nietzsche emphasizes, it is the craven desire to escape the burden of finding our way in the concrete and messy and terrifying business of the real world that drives the philosopher to invent abstract portraits of the "one true world" that sets a standard for life as against life (Nietzsche 1959). Our station is instead to affirm life, to achieve the greatness appropriate to us. By starting where we are in the concrete, this project assumes the hardest and greatest responsibility of creation and attention. "The salient difference between acting from a script and improvising is that one has to be not less but far *more* keenly attentive to what is given by the other actors and by the situation" (Nussbaum 1990, 94). This task of creation unfolds from the inside out, while embracing what is good and beautiful and true throughout the whole unfolding of the universe. And that is the ultimate promise of virtue ethics in its contemplation of Nature: "We learn that man has access to the entire mind of the Creator, is himself the creator in the finite. This view, which admonishes me where the sources of wisdom and power lie, and points to virtue as to 'The golden key; Which opes the palace of eternity,' carries upon its face the highest certificate of truth, because it animates me to create my own world through the purification of my soul" (Emerson 1982b, 73). The point of view of the universe may be suitable for a God or a beast, but not for a man, and we must not set before ourselves the project of abandoning our posts in a vain wish to be other than we are, for everything that matters that can be affirmed from the point of view of the universe can be equally seen and better appreciated from the point of view of the great soul. "Build therefore your own world" (Emerson 1982b, 81).

References

Aristotle. 2009. *Nicomachean Ethics.* In *Internet Classics Archive*, translated by W. D. Ross. http://classics.mit.edu/Aristotle/ (last accessed 10 Jan. 2009).

Brey, Philip. 2008. "Do We Have Moral Duties Towards Information Objects?" *Ethics and Information Technology* 10:109–114.

Dennett, Daniel. 1992. "The Self as a Center of Narrative Gravity." In *Self and Consciousness: Multiple Perspectives*, edited by Frank S. Kessel, Pamela M. Cole, and Dale L. Johnson, 103–15. Hillsdale, N.J.: Erlbaum.

Dostoevsky, Fyodor. 2010. *The Brothers Karamazov.* In *Fyodor Dostoevsky*, translated by Constance Garnett. http://fyodordostoevsky. com/etexts/the_brothers_karamazov.txt (last accessed 5 Jan. 2010). (Originally published in 1879.)

Emerson, Ralph Waldo. 1982a. "Circles." In *Selected Essays*, edited by Larzer Ziff, 225–38. New York: Penguin. (Originally published in 1841.)

———. 1982b. "Nature." In *Selected Essays*, edited by Larzer Ziff, 35–82. New York: Penguin. (Originally published in 1836.)

———. 1982c. "Self-Reliance." In *Selected Essays*, edited by Larzer Ziff, 175–205. New York: Penguin. (Originally published in 1841.)

Floridi, Luciano. 1999. "Information Ethics: On the Philosophical Foundation of Computer Ethics." *Ethics and Information Technology* 1, no. 1:37–56.

———. 2002. "On the Intrinsic Value of Information Objects and the Infosphere." *Ethics and Information Technology* 4, no. 4:287–304.

———. 2006. "Information Ethics, Its Nature and Scope." *Computers and Society* 36, no. 3 (September): 21–36.

———. 2007. "Global Information Ethics: The Importance of Being Earnest." *International Journal of Technology and Human Interaction* 3, no. 3:1–11.

———. 2008. "Information Ethics: A Reappraisal." *Ethics and Information Technology* 10:189–204.

Floridi, Luciano, and J. W. Sanders. 2004. "On the Morality of Artificial Agents." *Minds and Machines* 14, no. 3:349–79.

———. 2005. "Internet Ethics: The Constructivist Values of Homo Poieticus." In *The Impact of the Internet on Our Moral Lives*, edited by Robert Cavalier, 195–214. New York: State University of New York Press.

Foot, Philippa. 1988. "Utilitarianism and the Virtues." In *Consequentialism and Its Critics*, edited by Samuel Scheffler, 224–42. Oxford: Oxford University Press.

Hayek, Friedrich. 1958. "The Creative Powers of a Free Civilization." In *Essays on Individuality*, edited by Felix Morely, 77–84. Philadelphia: University of Pennsylvania Press.

Hunt, Lester. 1999. "Flourishing Egoism." *Social Philosophy and Policy* 16, no. 1:72–95.

Nagel, Thomas. 1986. *The View from Nowhere*. Oxford: Oxford University Press.

Nietzsche, Friedrich. 1959. *The Twilight of the Idols*. In *The Portable Nietzsche*, edited and translated by Walter Kaufmann. New York: Viking Penguin. (Originally published in 1899.)

Nussbaum, Martha. 1990. "The Discernment of Perception: An Aristotelian Conception of Private and Public Rationality." In her *Love's Knowledge*, 54–106. Oxford: Oxford University Press.

———. 1995. "Aristotle on Human Nature, and the Foundation of Ethics." In *World, Mind, and Ethics*, edited by J. E. J. Altham and Ross Harrison, 86–131. Cambridge: Cambridge University Press.

Popper, Karl. 1985. "Historicism." In *Popper Selections*, edited by David Miller, 289–303. Princeton: Princeton University Press. (Originally published in 1936.)

Rawls, John. 2005. *Political Liberalism*. New York: Columbia University Press.

Stahl, Bernd. 2008. "Discourses on Information Ethics: The Claim to Universality." *Ethics and Information Technology* 10:97–108.

Thiele, Leslie Paul. 1990. *Friedrich Nietzsche and the Politics of the Soul: A Study of Heroic Individualism*. Princeton: Princeton University Press.

Williams, Bernard. 1985. *Ethics and the Limits of Philosophy*. Cambridge, Mass.: Harvard University Press.

———. 1988. "Consequentialism and Integrity." In *Consequentialism and Its Critics*, edited by Samuel Scheffler, 20–50. Oxford: Oxford University Press.

———. 1994. "Must a Concern for the Environment Be Centred on Human Beings?" In *Reflecting on Nature*, edited by Lori Gruen and Dale Jamieson, 46–52. Oxford: Oxford University Press.

10

THE PHILOSOPHY OF INFORMATION:
TEN YEARS LATER

LUCIANO FLORIDI

This is a collection of very fine chapters. Their scope, depth, and insightful-ness are testimonies not only to the brilliance and scholarship of their authors but also to the remarkable maturity reached by the philosophy of information (PI) during the past decade. In the late 1990s, I was searching for an approach to some key philosophical questions: the nature of knowledge, the structure of reality, the uniqueness of human conscious-ness, a satisfactory way of dealing with the new ethical challenges posed by information and communication technologies, to list some of the topics discussed in this collection. I had in mind a way of doing philosophy that could be rigorous, rational, and conversant with our scientific knowledge, in line with the best examples set by the analytic tradition; non-psycholo-gistic, in a Fregean sense; capable of dealing with contemporary and lively issues about which we really care; and less prone to metaphysical armchair speculations and idiosyncratic intuitions. I was looking for a constructive philosophy that would provide answers, not just analyses, that would be as free as possible from a self-indulgent, anthropocentric obsession with us and our super-duper role in the whole universe, and respectfully sceptical of commonsensical introspections and Indo-European linguistic biases. It was a recipe for disaster, but then, sometimes, fortune favours the irresponsible. During that period of intellectual struggle and confusion, I realised one day that what I had in mind was really quite simple: a philosophy grounded on the concept of information. I was not on my way to Damascus but in Oxford, at Wolfson College, sitting on the bank of the river Cherwell, when I discovered that the spectacles I was looking for were on my nose. It was the summer of 1998. Six months later, I gave a talk in London, at King's College, entitled "Should There Be a Philosophy of Information?" The question was of course rhetorical, and I soon started working on the essay that became "What Is the Philosophy of Information?" It was published in *Metaphilosophy* in 2002 (Floridi 2002).

Almost a decade later, it is reassuring to see that the project for a philosophy of information as a discipline in its own right was not ill conceived. Witness the fact that it would be impossible to do full justice both to the quality of the contributions in this collection and to the value of the new and exciting area of research to which they belong. For this reason, in the following pages I shall not try to summarise or discuss every interesting issue raised by the contributors. Rather, I shall briefly highlight and seek to clarify some critical points, with the goal of at least reducing our disagreement, if not achieving a full convergence of views. As I remarked in a comparable context (Floridi 2008a), philosophy deals with problems that are intrinsically open. Intellectual disagreement is therefore an essential part of its conceptual explorations. When it is informed and reasonable, it should be welcome, not eradicated. It is a very sad restaurant, soon to be out of business, in which you can order only the same dish, no matter how delicious it is. At the same time, the controversies contained in this collection should not eclipse the fact that there is a great deal about which we all agree, in terms of importance of topics, priority of problems, and choice of the best methods to be employed to address them. We would not be engaged in this lively dialogue if we did not share so much intellectually.

Let me now add a final word before closing this introductory section. There will be no space below to summarise the main lines of my research. So the reader interested in knowing more about my work might wish to have a look first at a very gentle introduction to the nature of information, written for the educated public (Floridi 2010a). The more adventurous reader might be interested in knowing that most of the essays referred to in this collection were part of an overarching project and have now found their proper place, as revised chapters, in a single, more technical book (Floridi 2010b).

Comments and Replies

I shall be extremely brief in my comments on Hendricks's and Roush's interesting chapters. Hendricks shows how *minimalism* might be useful in tackling the problem of "pluralistic ignorance," where ignorance is to be understood as lack of information, rather than lack of knowledge. I might have overlooked some error, but I must confess that I fully agree with both the analysis and the proposed solution. Roush follows a similar pattern, although in her case it is the *knowledge game* that she finds valuable in order to approach the "swamping problem," that is, the question of what added value knowledge might have, over and above mere true belief, or, I would add, mere information. What both chapters share, methodologically, is a careful approach to formal details; a problem-solving orientation that allows the selection of the right information-theoretical tools; and a treatment of informational agents, their environments and the issuing processes free from psychologistic features. This is where PI most fruitfully joins forces with recent trends in formal epistemology. In both chapters, for example, the agents involved could be companies, political parties, or individual human beings; it does not matter. In terms of contents, there is one more feature that I would like to stress: the central role played by *equilibria*, both negatively, when in Hendricks's chapter we need to disrupt pluralistic ignorance, and positively, when in Roush's chapter we need to discover what stable conditions lead to an optimal epistemic relation with the world. Impeccable.

The next chapter, by Bringsjord, also focuses on the *knowledge game* (KG). As often when reading Bringsjord, my temptation is to treasure his insights and keep quiet. Given the circumstances, I will have to resist it. So let me start by saying that Bringsjord's mastery of KG is not only flawless but impressive. I might have invented the game, but he certainly knows how to play it elegantly. Bringsjord and I agree that—given the *current* state and understanding of computer science—the best artefacts that artificial intelligence (AI) will be capable of engineering will be, at most, zombies: artificial agents capable of imitating an increasing number of human behaviours. In most cases (I am sure Bringsjord agrees), such agents will also be better than the fragile and fallible humans who provide the original templates. In some cases—here Bringsjord might disagree, but see below—human capacities will remain unmatched. I have in mind primarily our semantic abilities. In any case, and for a variety of converging reasons, neither of us is convinced that human-like minds might be engineered artificially. We also agree that current, off-the-shelf, artificial agents as we know them nowadays cannot answer self-answering questions and pass the test.

Where we seem to part company is in deciding whether this holds true also for *foreseeable* artificial agents. I think so, but Bringsjord offers a proof that this is not the case. In other words, he can foresee and forecast artificial agents that will pass the KG test. I am happy to concede the point. He might be right. Or maybe not. *Partly*, it is a matter of *details*, which Bringsjord, absolutely rightly, could not fully provide in his chapter. Not his fault, but they remain the preferred hiding places for devilish disappointments. *Partly*, it is a matter of *implementation*. Sometimes what looks plausible on paper turns out to be unfeasible on the ground, thus proving to be only a logical and not an empirical possibility. *Partly*, it is a matter of *interpretation*. Passing a test means being able to pass it regularly and consistently, according to qualified judges, not occasionally or only thanks to a favourable setting. Think of a reading test: if you could read only sometimes, and only when looking at a text you previously memorised, you would not qualify as a reader. Now, Bringsjord acknowledges that "current logic-based AI is able to handle some self-answering questions. Notice, though, that I say 'some' self-answering questions. There is indeed a currently insurmountable obstacle facing logic-based AI that is related, at least marginally, to self-answering questions: it is simply that current AI, and indeed even *foreseeable* AI, is undeniably flat-out impotent in the face of *any arbitrary* natural-language question—whether or not that question is self-answering." So it seems that my cautious attitude is vindicated. However, let us forget about all these inclusions of "partly." Let us assume, for a moment, that all possible reservations turn out to be *de facto* unjustified: details are provided, the implementation works, and there is no hermeneutic disagreement about what is going on. Test passed. This is the real point at which I may not be able to follow Bringsjord any further. For, as I wrote in "Consciousness, Agents and the Knowledge Game," the KG was never meant to provide a "Floridi challenge" for AI. Let me quote the relevant passage (emphasis added):

> What is logically possible for (or achievable at some distant time in the future by) a single articial agent or for an articial multiagent system *is not in question*. I explained at the outset that we are not assuming some science ction scenario. "Never" is a long time, and I would not like to commit myself to any statement like *"articial agents will never be able (or are in principle unable) to answer self-answering questions"*. *The knowledge game cannot be used to argue that AI or AC (articial consciousness) is impossible in principle*. In particular, its present format cannot be used to answer the question "How do you know you are not a futuristic, intelligent and conscious articial agent of the kind envisaged in Blade Runner or Natural City?" As far as articial agents are concerned, the knowledge game is a test to check whether AI and AC have been achieved.
>
> (Floridi 2005a, 431)

So much so that the article ends by showing a slightly puzzling result: the test can also be passed by a multi-agent system made of zombies. This is one more reason why I would not be surprised if Bringsjord were completely right. A challenge is usually a negative modal statement of the form (c) "*x* cannot ϕ," for example, "John cannot run the marathon." A test is usually a conditional (possibly modal) statement of the form (t) "if *x* (can) ϕ then *x* qualifies as *y*," for example, "if John (can) runs the marathon then John is fit." It is true that the same ϕ, for example, running the marathon, enables one to meet the challenge and pass the test. This is where Bringsjord's and my line of reasoning run parallel. But then, he seems to believe that I am arguing in favour of a specific interpretation of (c), whereas I am interested in (t). The purpose of the KG is that of providing a test whereby you may show, given that you are conscious (not a zombie), how you know that you are (not a zombie). If that is achieved, that was the challenge I wished to meet.

The chapter by Scarantino and Piccinini and the chapter by Adams form a perfect diptych in which the former provide some criticisms that are well answered by the latter. Adams himself, of course, has his own reservations and, as we shall see presently, they are to be taken seriously, but let me comment first on Scarantino and Piccinini's contribution.

Scarantino and Piccinini are right in stressing the need for a *pluralist approach* to the many different uses of "information." If I may phrase such pluralism in my own words: "Information 'can be said in many ways,' just as being can (Aristotle, *Metaphysics* Γ.2), and the correlation is probably not accidental. Information, with its cognate concepts like computation, data, communication, etc., plays a key role in the ways we have come to understand, model, and transform reality. Quite naturally, information has adapted to some of Being's contours. Because information is a multifaceted and polyvalent concept, the question 'what is information?' is misleadingly simple, exactly like 'what is being?'" (Floridi 2003, 40). As a consequence, any thesis about the nature of information, including the one about the *veridicality* of *semantic* information, should be handled with extra care. It would be daft, for example, to identify a piece of software as information—as we ordinarily do in IT and computer science—and then argue that, since information must be true, so must be that piece of software. "True about what?" would be the right sceptical question. Likewise, it would be unduly pedantic to insist that, given the veridicality thesis, cognitive scientists should stop speaking about information processes. Sometimes, they may be talking about information in a non-semantic sense; some other times, they may just be using a familiar synecdoche, in which the part (semantic information) stands for the whole (semantic information and misinformation), as when we speak in logic of the truth-value of a formula, really meaning its truth or falsehood. Often, they are using information as synonymous for data, or representations, or contents, or signals, or messages, or neurophysiologic patterns, depending on the context, without any loss of clarity or precision. The reader looking for an initial map of the varieties of senses in which we speak of "information" can find it in Floridi 2010a.

Since I agree that information "can be said in many ways," I also subscribe wholeheartedly to Scarantino and Piccinini's invitation to adopt a *tolerant attitude* towards the uses to which the concept of information can be put. Bananas are not fruit, and tomatoes are not vegetables, but we know where to find them in the supermarket, and not even a philosopher should complain about their taxonomically wrong locations. So why, given their pluralism and tolerance, are Scarantino and Piccinini so keen on rejecting the veridicality thesis?

The thesis in itself seems to be fairly harmless and most reasonable (Floridi 2005b, 2007). When you ask for some semantic information—about when the supermarket is open, for example—you never specify that you wish to receive *truthful* information. That goes without saying because that is what semantic information is. So if you get "false information" and go to the supermarket when it is closed, you may rightly complain about your source for having provided you with no information at all. Some semantic content c qualifies as semantic information i only if c is truthful. If it is not, then c is misinformation at best, disinformation at worst (this being misinformation wilfully disseminated for malicious purposes). Simple.

The veridicality thesis is also hardly original. It has been treated as obvious by several philosophers who have handled semantic information with all the required care, including Grice, Dretske, and Adams. It does make life easier when dealing with difficult and controversial issues such as some paradoxes about the alleged informativeness of contradictions (they are not informative now because they are false [Floridi 2004]); the link between semantic information and knowledge (knowledge encapsulates truth because it encapsulates semantic information, which, in turn, encapsulates truth, as in a three-doll matryoshka [Floridi 2006]); or the nature of relevant information [Floridi 2008b]). Despite all this, it would be ungenerous to dismiss the contribution by Scarantino and Piccinini as a fruitless mistake. Let me try to explain why.

Imagine that Mary is told by John that "the battery of the car is flat." This is the sort of semantic information that one needs to consider when arguing in favour of the veridicality thesis. The difficulty is that such semantic information is the result of a complex process of elaboration, which ends with truth but certainly does not start from it. Indeed, one of the great informational puzzles is how physical signals, transduced by the nervous system, give rise to high-level, truthful semantic information. When John sees the red light flashing, there is a chain of data-processing events that begins with an electromagnetic radiation in the environment, in the wavelength range of roughly 625–740 nanometres, goes through John's eyes and nervous system, is elaborated by him in terms of a red light flashing in front of him, is combined with regular associations on the physical side (the light being red is coupled to the battery being flat by the engineering of the system) as well as with background knowledge on John's side (e.g., concerning signals of malfunctioning in cars). It all ends with Mary receiving John's message that "the battery is flat." Some segments of this extraordinary journey are known (again see Floridi 2010a for a simple introduction to it), but large parts of it are still mysterious. Now, if one wishes to talk rather loosely of information from the beginning to the end of this journey and all the way through it, that is fine. We know our way in the supermarket, we can certainly handle loose talk about information. There is no need to be so fussy about words: the tomatoes will be found next to the salad, and the bananas next to the apples. So I am fully convinced by Scarantino and Piccinini: from such a "supermarket approach," the veridicality thesis is untenable, since truth or falsehood plays absolutely no role in "information" for a long while during the journey from electromagnetic radiation to "Sorry, dear, the battery is flat." Of course, this leaves open the option of being conceptually careful when dealing with semantic information itself, the end product of the whole process. Botanically, tomatoes and courgettes are fruit, and bananas are female flowers of a giant herbaceous plant.

Likewise, in philosophy of information semantic information is well formed, meaningful and truthful data. If you still find the veridicality thesis as counterintuitive as the fruity tomatoes, just assume that Grice, Dretske, Sequoiah-Grayson, Adams, I and anyone else who endorses it are being botanically precise and talking about *premium semantic information*. As we saw above, some pluralism and tolerance will help.

A final comment before turning to Adams's chapter. I believe Scarantino and Piccinini might be on to something interesting, namely, a project concerning the various uses and meanings of information in cognitive science. If I am not mistaken, then this is most welcome, as their investigations will provide a much-needed insight into an area still under-investigated. Of course, it is to be hoped that such a project will complement and build upon the fundamental research by Barwise and Seligman on the logic of distributed systems and the analysis of information flow, and hence be consistent with their results, including the fact that "the proposal agrees with Dretske's in that it makes information veridical. That is, if a is of type α and this carries the information that b is of type β, then b is of type β" (Barwise and Seligman 1997, 36). I shall return to this point in a moment, in connection with Adams's rejection of the *distributive thesis* (information closure).

Adams's elegant and insightful chapter deserves to be studied carefully. There is much about which I completely agree, and more that I have learnt. When I wrote above that Adams provides a useful answer to issues raised by Scaratino and Piccinini, I had in mind things like his clear and correct distinction between "a semantic notion of information . . . [understood as] information that p or about state of affairs f that exists in one's cognitive system (one's beliefs, or perceptions or knowledge)" and "information in the sense of natural sign or nomic regularity, where information can exist outside cognitive agents." This is only an example, and the chapter merits close analysis. Here I shall deal only with Adams's two objections.

The first concerns the possibility of having a system acquire some semantics (ground at least some of its symbols) through supervision. According to Adams, helping a system (whether human or artificial, it does not matter) to acquire the meaning of a symbol s might be a case of *causal derivation*, not of what I would call semantic cheating, that is, a case in which the trainer has the meaning pre-packaged and transfers it to the trainee, who not only would have been unable to acquire it except for such a present but above all still does not master it. I must admit that I remain unconvinced. We know that artificial systems with semantic capabilities are not yet available. Leave to one side whether they will ever be. The fact that currently they are not shows that the symbol-grounding problem remains unresolved. And I suspect that this is so because causally inducing a system to behave as if it had acquired and mastered the meaning of a symbol s is useful but still insufficient to guarantee that that system actually has gained a semantics for s and the capacity to use it efficiently. I agree with Adams that causal derivation might be *sufficient to teach* a potentially semantic system (e.g., a dog or a human being) the meaning of s, given the right circumstances (but see below the case of the slave boy). I also agree that the same causal derivation might be *sufficient to transmit* the meaning of s, from a system that enjoys its semantics to a system that might acquire it. I even agree on the fact that causal derivation might play a *role, perhaps crucial, in creating* the meaning of s, as when "nature teaches" a system the meaning of s. Where Adams and I might (but I am not sure) disagree is that causal derivation is sufficient in generating the meaning of s ex nihilo, without presupposing any previous semantics of s or another system proficient in handling its meaning. The problem is as old as Plato: does the slave boy (the close equivalent to a robot or "animated instrument")[1] really understand that the diagonal of any square is the base of a square of double the area? Or is it only because Socrates knows Pythagoras's theorem that he is able to induce the slave boy, gently but firmly, through causal derivation, to say what Socrates wishes him to say, like a dumb but well-trained parrot? I am sceptical about the slave boy's actual acquisition of the right semantics. But even if you do not share my scepticism, and side with Plato, note that ultimately the Platonic solution is to make the slave boy's semantics innate. Either way—the semantics is not really there or the semantics has been there all along, because pre-implanted—I remain unconvinced by Adams's position and his comments regarding the zero-semantic commitment (see his footnote 3).

The second objection is a different case. I believe here Adams is right, and that he is so in an important sense. He argues that we should reject information closure, that is, the *distributive thesis* according to which, if a is informed both that p and that $p \rightarrow q$ then a is also informed that q. As Patrick Allo kindly reminded me, in general the problem is (at least) twofold (see Allo forthcoming). One might wish to reject information closure

(i) either because a may not be informed that p;
(ii) or because p does not count as semantic information;
(iii) or both, of course.

---

[1] "*So a slave is an animated instrument*, but every one that can minister of himself is more valuable than any other instrument; for if every instrument, at command, or from a preconception of its master's will, could accomplish its work (as the story goes of the statues of Daedalus; or what the poet [Homer] tells us of the tripods of Vulcan, 'that they moved of their own accord into the assembly of the gods'), the shuttle would then weave, and the lyre play of itself; nor would the architect want servants, or the [1254a] master slaves." Aristotle, *Politics*, book I, chapter V, in *The Politics of Aristotle: A Treatise on Government*, trans. William Ellis (London: J. M. Dent and Sons, 1912).

I understand Adams as being concerned only with (i) in his contribution. His support of a tracking theory of knowledge, his references to Dretske, and his examples involving Chris all point in this direction unambiguously. So I shall dedicate much more attention to (i). On (ii) (and hence [iii]), let me just add that it is sufficiently open to interpretation to allow different views on the value of information closure, but that Adams and I appear to agree on the following. Recall the quotation above from Barwise and Seligman 1997. Suppose two systems a and b are coupled in such a way that a's being (of type, or in state) F is correlated to b's being (of type, or in state) G, so that $F(a)$ carries (for the observer of a) the information that $G(b)$. Information closure in this case means that, if $F(a) \rightarrow G(b)$ qualifies as information and so does $F(a)$, then $G(b)$ qualifies as well: if the low-battery indicator (a) flashing (F) indicates that the battery (b) is flat (G) qualifies as information, and if the battery indicator flashing also counts as information, then so does the battery being flat. Where Adams and I might disagree (but see below) is in relation to (i). As Adams acknowledges, I reject the *distributive thesis* in cases in which the kind of information in question is empirical (seeing, hearing, and so on), but not when it is semantic. He would like to see a more uniform approach, I resist it, but we might not be at variance. Consider the following case.

In the left pocket of your jacket you hold the information that, if it is Sunday, then the supermarket is closed. Your watch indicates that today is Sunday. Do you hold the information that the supermarket is closed today? The unexciting answer is maybe. Perhaps, as a *matter of fact*, you do not, so Adams is right. You might fail to make the note in the pocket and the date on the watch "click." Nevertheless, I would like to argue that you should, that is, that as a matter of *normative analysis*, you did have the information that the supermarket was closed. So much so that you will feel silly when you are in front of its closed doors and realise that, if you had been more careful, you had all the information necessary to save you the trip. You should have known better, as the phrase goes. Now, I take logic to be a normative discipline. From this perspective, the *distributive thesis* seems to me to be perfectly fine. Still from the same perspective, the *distributive thesis* is not always applicable to empirical information. Adams is talking about the performance of actual players, I am talking about the rules of the game. Consider the same example. This time you *read* the following e-mail, sent by the supermarket: "The shop will be closed every Sunday." You also read the date on your computer, which correctly indicates that today is Sunday. Have you read that the supermarket is closed today? Of course not, as we assume that there were no further messages. Should you have read that it was? Obviously not, for where was the text that you should have read? Should you have inferred that the supermarket was closed today? Surely, for that was the information that could easily be inferred from the two texts that you read. If Adams's thesis is that information closure is at best only a matter of normative logic and certainly not an empirical fact, I am convinced.

Like Adams's, Colburn and Shute's chapter is another contribution from which I have learnt much. I believe Colburn, Shute and I fully converge on roughly the same conclusions, even if coming from rather different perspectives. I find this most reassuring, as evidence of a robust and convincing indication of a sound methodology. In light of this agreement, I would like to take this opportunity to stress two aspects of the method of levels of abstraction (LoAs).

First, it is true that once it is imported into a philosophical context, it is harder to re-apply the method in computer science in precisely the same new format. Some of the exact formalisms are inevitably lost, in order to make room for conceptual and qualitative flexibility, and this justifies the fact that it is a one-way adaptation. However, computer science must be credited for providing the conceptual resources that have led to the philosophical approach based on LoAs. We can do better philosophy by learning such an important lesson from our colleagues in the computer science department, and this much is to be acknowledged. It is a matter not of pedigree but of giving to Turing what is Turing's.

The other aspect concerns exactly the roots of the method of abstraction in computer science, a science that, in so far as it is also a branch of engineering, builds and modifies its own objects, exactly as economics, law, and the social sciences may build and modify social life and interactions. LoAs do not represent only a hermeneutic device, like Dennett's stances. They are the conditions of possibility of informational access to systems and hence what determine our models of them. In this sense, they have an ontic value that stances or other forms of "perspectivism" cannot but lack.

Colburn and Shute's chapter ends with some considerations on the possibility of an ontology based on informational structures. This is where the thread of the conversation is picked up by Bueno's chapter. His contribution is a welcome expansion in the number of options favourable to some form of structural realism. It seems to contain, however, just a couple of unfortunate misunderstandings of my position, which might be worth dissipating here in order to leave the reader with a set of clearer alternatives.

Bueno seems to accept the veridicality thesis. Excellent news. However, he is a bit cautious in some cases. He need not be. Everyone defending the veridicality thesis would agree with him when he writes that "information can be used successfully, but it need not be true for it to play a successful role. Truth is not required for empirical success, not even novel empirical success involved in the discovery of a new planet," for the following reason: "It is easy to be confused about both 'relevance' and 'misinformation' ['false information']. . . . That misinformation may turn out to be useful in some serendipitous way is also a red herring. False (counterfeit) banknotes may be used to buy some goods, but they would not, for this reason, qualify as legal tender. Likewise, astrological data may, accidentally, lead to a scientic discovery but they are not, for this reason, epistemically relevant information. Of course, there are many ways in which misinformation may be indirectly, inferentially or metatheoretically relevant, yet this is not what is in question here" (Floridi 2008b, 84–85).

Bueno speaks of Bode's law instead of astrological data, but the reader can see where the problem lies: understanding the veridicality thesis as if it were inconsistent with the usefulness of false content (misinformation) would be a mistake, not least because the usefulness of misinformation can be presented as a case of *ex falso quodlibet*. In a different context (Floridi 2005b), I also explained why some misinformation—for example, to be told that there will be three guests for dinner tonight when in fact only a couple is coming—might be much preferable to vacuous semantic information—for example, to be told that there will be fewer than a hundred guests tonight. Again, it is a simple exercise left to the reader to draw a similar conclusion about Newtonian physics. All this is really not a serious issue, and could be disposed of as just marginal details that can easily be rectified. Much more interesting is to try to understand what sort of theory of truth would allow us a robust approach to strictly speaking false but still valuable theories (or semantic misinformation). If I am told that the train leaves at 12.15 when in fact it leaves at 12.25, I might be slightly annoyed, but I can still catch it. In other words, an informee should prefer semantic information, but, short of that, she could still settle for, and exercise plenty of tolerance towards, valuable misinformation, that is, false content that still allows her to interact with the targeted system successfully. I must confess that I am not keen on "quasi-truth," since it seems to me to be a label under which one might hide the difficulty rather than clarify it. But I take the problem as seriously as Bueno, and I have tried to solve it through a correctness theory of truth that might deal with such cases (Floridi forthcoming 2). Even a short outline of it would take us too far away here, but let me just say that the basic idea is to analyse truth in terms of correctness, and correctness in terms of a commutative relation (in the category theory's sense of "commutation") between the model under discussion and its target system, that is, between the proximal access to some semantic information and the distal access to its target system.

It seems that a correctness theory of truth places Informational Structural Realism (ISR), a bit closer to Bueno's Structural Empiricism, yet Bueno is concerned that I might actually be a sceptic. Allow me to put myself in excellent company: this was, and still is sometimes, the same accusation moved against Kant. He who denies epistemic access to the ultimate nature of things in themselves must be a friend of Sextus Empiricus. Absolutely not. Scepticism, when properly understood, is a family of philosophical arguments in favour of the impossibility of establishing, with total certainty, whether we have reached the truth about some particular matter. Translated into informational terms, it is an attack against the possibility of determining whether some semantic content c about a target system s might actually be a case of semantic information i about s. Is $c_s = i_s$? The sceptic does not argue in favour of a negative answer but seeks to show that one can never tell. ISR, on the contrary, is in favour of the possibility of answering an endless number of occurrences of such type question, although from a fallibilist position of course, since we might be, and have been, wrong in the past (see Floridi 1996 and forthcoming). So the real divide is not between my sceptical and Bueno's anti-sceptical position but a constructionist understanding of knowledge, which is essential to grasp ISR but which Bueno disregards. This is not surprising, since Bueno seems to favour some representation-alist theory of information/knowledge, which I do not endorse. To put it very simply, I support a maker's knowledge approach (Bacon, Hobbes, Vico, Kant, Cassirer) and hold that gaining information and hence knowledge about the world is a matter of data *processing*, where "processing" is taken very seriously. As with cooking, the end result of our cognitive (including scientific) elaborations is absolutely realistic, since without ingredients and proper baking there is no cake; but the outcome does not represent or portray, or x-morph, or take a picture of the ingredients or of the baking. Knowledge delivers conceptual artefacts, which are as real and objective as the cake you are eating. This anti-representationalism should not be confused with any version of anti-realism.

A final remark before closing. So far, I have been talking as if ISR concerned information only understood semantically, epistemically, cog-nitively, or methodologically. It does not. ISR defends primarily an *ontological* thesis, namely, an informational understanding of reality. This seems a significant difference from Bueno's structural empiricism. It is well captured by Steven French (forthcoming):

> In effect what Floridi's approach allows us to do is separate out the commitments of ESR [Epistemic Structural Realism] from both Worrall's agnosticism and Poincaré's espousal of 'hidden natures'. At the first order LoA [level of abstraction], ESR—as the name suggests—offers us an epistemic form of realism to the effect that what we know in science are the relevant structures (and if we are to follow this line of analysis, we should perhaps drop or at least modify the afore-mentioned slogan). Beyond that, Poincaré, Worrall *et al.* should remain quiet; there should be no talk of natures, hidden or otherwise, no adoption of forms of agnosticism, but rather a 'quietist' attitude to any further commitments. To make those is to proceed to the next level, as it were,

and here the appropriately metaphysically minimal attitude is that offered by OSR [Ontic Structural Realism], which reduces the amount of humility we have to swallow by reconceptualising the underlying (putative) objects themselves in structural terms.

In a popularization of his views on the ultimate nature of reality, Frank Wilczek (2008) presents ordinary matter as a secondary manifestation of what he calls *the Grid*, namely (what we perceive as mere) empty space (but that is actually) a highly structured entity. Wilczek was awarded the Nobel Prize in physics in 2004 for his contribution to the discovery of asymptotic freedom in the theory of the strong interaction. I shall not pretend to understand the sophisticated physics that led him to his views about matter. Metaphysically, however, I am very sympathetic, because it seems to me that Wilczek's Grid is the physicist's counterpart of what I have defined as the *infosphere*: a non-materialist view of the world as the totality of informational structures dynamically interacting with each other. This is the ontology I defend in ISR.

The final chapter on which I shall comment looks at my work on the ethics of the infosphere. Volkman's main contention is clearly stated at the beginning of his contribution: information ethics (IE) is too foundational, impartial, and universal to "do full justice to the rich data of ethical experience." The use of "data" might have been a Freudian slip, but the point is unmistakable: IE is partly useless, partly pernicious. Reading the chapter, one has the impression of a Manichean dichotomy between two moral discourses: the good one, which is warm, human, careful about the richness of life and the complexity of our difficult choices, bottom-up, with roots in real cases and full of *phronesis*; and the bad one, which is cold, objectifying, abstract, unable to capture the nuances of everyday experience, top-down, detached and algorithmically calculating. Virtue ethics (VE) versus IE, in case you had any doubts. If only things were so simple. The outcome of such a Manichean view is that Volkman's chapter contains many insightful remarks, but very few of them concern IE. Anyone interested in an informed and reasonable discussion of IE might prefer reading the chapter by Terry Bynum that provides the Epilogue to this collection, or the excellent essay by Charles Ess (2008).

The problem with Volkman's approach is that it seeks to build a conflict of views at the cost of unfairness and lack of objectivity, when a more constructive and fruitful dialogue between IE and VE could have identified many convergences, as well as potentially bridgeable disagreements and complementary divisions of interest, thus acknowledging what each theory might be better positioned to provide. Personally, I have often argued that the distance between IE and VE is small, since both call our attention to the need to develop morally good constructions of agents and their societies and of the natural and artificial environments. *Poiesis* is a fundamental activity that requires careful ethical investigations, and it is fair to claim that it is IE that has stressed its crucial ethical importance in the current debate (cf. the concept of *homo poieticus*). The increasing interest in the ethics of design is proof of such timely focus. But this is how far Volkman is prepared to be friendly towards IE. Having grasped this point, he adopts the Manichean dichotomy illustrated above and tries, unsuccessfully though strenuously, to transform differences in focus, emphasis, and scope into a deep and irrecoverable fracture.

Interest in the chapter as a critical discussion of IE starts waning once it charges IE with the patently impossible pretence of "incorrectly suppos[ing] that there are judgments regarding the being and flourishing of information entities that are not bound to the perspective of some agent, and that these judgments can enter into human decisions about what to do and who to be." This straw man, the cold and dry view from the sky that Volkman is keen to slap onto IE, is nowhere to be found in my or indeed other colleagues' work on IE. Not least because, as Volkman acknowledges, IE firmly holds that ethical investigations must be developed by adopting and specifying the levels of abstraction (LoA) at which they are conducted, and therefore the context and purposes for which a LoA is privileged. If the reader is put off by "levels of abstractions," as something that sounds too close to a cold logical formalism, let me suggest replacing them here with warmer "human views." What IE argues is that our intrinsic animal biases, our egocentric drives, and our anthropocentric inclinations can be withstood, mitigated, and rectified, through reflection, education, social pressure, and a progressive improvement in our understanding of our roles in the universe. We start as selfish egoists interested only in ourselves, Hobbes is right, but we can and must hope to become unselfish and altruistic stewards of the world. Failure along the way is inevitable, especially at the individual level, but whatever small degree of success is achieved, it should be most welcome. We can become better agents by progressively balancing the demands of the shouting "me, always me, only me!" with the demands of the other, both biological and artificial. This is why, for example, IE is regularly compared to Buddhist ethics. Of course, we must "start where we are," as Volkman repeatedly recommends. Yet this is trivial. There is no other place where we could start. The interesting question is whether staying where we accidently find ourselves thrown by natural evolution is good enough. IE argues that it is not (Hongladarom 2008). The alternative is an ego-colonialism that is unappealing. We read in the chapter that "although I cannot succeed in my life by becoming someone else, it is equally true that my own success depends on extending my self by including others in my very constitution." It is on this well-meant inclusiveness from within, rather than respectful acceptance from without, that some of the worst deeds have been justified. Especially nowadays, it seems irresponsibly self-indulgent to enjoy the reassuring scenarios in which there are only friends and loving agents in ethics, while the rest is politics. What happens when the world neither wishes nor consents to be included in our "very constitution" but asks respectfully to remain other from us? How can we deal with conflicts between polarised agents, all bent on "starting where they are" and unwilling to step out of their egocentric predicaments? IE rejects the option to "Go out into the highways and hedges and force them to enter that my house may be filled" (Luke 14:23). Augustine was keen on that passage, which provided textual justification for the Crusades.

Unfortunately, having made the crucial false step of misunderstanding IE for a cold, objectifying, abstract approach to human morality, the chapter stumbles on several other points. The contemporary shift of the ethical discourse, from being entirely agent-centred to being progressively (and at least equally if not) more patient-centred is disregarded at a cost, although it represents a crucial novelty in such areas as medical ethics, bioethics, environmental ethics, or indeed information ethics. Accusations of *historicism* (or, alternatively, *anachronism*, if the historical development fails to support a theory) leave the conceptual debate untouched. The list of other missed opportunities to debate IE in its real nature rather than as a caricature is too long not to become tedious. For example, pluralism is intrinsic to IE, which also defends the crucial importance of the *overridable* nature of the respect to be paid to informational entities, a feature that explicitly makes IE both willing and able "to discriminate between the information entities that merit respect and admiration and those that have not earned this status." Or take Volkman's misrepresentation of the boy in the junkyard example. I provided it as simplified thought experiment to illustrate pros and cons of different ethical theories. Volkman uses it as a target to which he addresses rhetorical questions: "Is the boy really just getting mindless kicks, or is he rehearsing his shot? How much time are we talking about? What are the alternatives open to him? What brought him here, and where is he going? There are myriad coherent stories in which it would be perfectly O.K. to smash things." But the rhetorical game of adding "richness" to an intentionally streamlined example is trivial, and anyone can play it: "Nobody grants that breaking windscreens necessarily leads to a bad character, life is too short to care and, moreover, a boy who has never broken a car windscreen might not become a better person after all, but a repressed maniac, who knows? Where did David practice before killing Goliath? Besides, the context is clearly described as ludic, and one needs to be a real wet blanket to reproach a boy who is enjoying himself enormously, and causing no apparent harm, just because there is a chance that his playful behaviour may perhaps, one day, slightly contribute to the possible development of a moral attitude that is not praiseworthy" (Floridi 1999, 54).

I fully subscribe to the view that "if impartialism and universalism turn out to be undesirable in themselves, at least when carried beyond their appropriate domains, then much of ethics since the Enlightenment has been a mistake, with IE as the most recent and most glaring example." It is exactly because I believe that much of ethics since the Enlightenment has been a success and that IE is the most recent development of such a worthy tradition that I wholeheartedly hope that ethics will maintain a reasonable defence of both impartiality and universality. A fair and tolerant society depends on them, and we are getting more global by the day. We need to find a way to dialogue impartially and universally. What one might argue is that the impartial and universal application of morality needs to be consistent with the diversity of the agents and patients involved, and the variety of their predicaments. This is not a point made by Volkman, but I doubt anyone would disagree about it.

In conclusion, the chapter represents a missed opportunity. Since it opens with a famous and beautiful quotation from Emerson, allow me to close my few remarks with a classic one by Shakespeare:

> HORATIO: O day and night, but this is wondrous strange!
> HAMLET: And therefore as a stranger give it welcome.
> There are more things in heaven and earth, Horatio,
> Than are dreamt of in your philosophy. (*Hamlet*, act 1, scene 5)

Conclusion

Information has been a subject of philosophical interest for a very long time. In a way, one could read the whole history of philosophy as containing a thin electric-blue line that runs from the pre-Socratic philosophers to us. Obvious developments in our technology, society and culture have brought to light such continuous, uninterrupted thread, which I have characterised in my work as the philosophy of information (PI). PI has opened up a very rich area of conceptual investigations. Now, the development of new philosophical ideas seems to be more akin to economic innovation than we usually acknowledge. For when Schumpeter (1943) adapted the idea of "creative destruction," in order to interpret economic innovation, he might as well have been talking about intellectual development. This is the way I understand the metaphor of the digital phoenix used by Bynum and Moor (1998) (see the next chapter). This collection shows how much creative destruction has been caused by PI. I hope it is only the beginning.

Acknowledgments

I am most grateful to all the contributors in this collection for their time, insightful comments and tough challenges; to Patrick Allo for his patient and excellent work as guest editor, and his insightful Introduction; and to Otto Bohlmann (managing editor) and Armen T. Marsoobian (editor in chief) of *Metaphilosophy* for having organised and hosted this under-taking and allowed us to work together.

References

Allo, Patrick. Forthcoming. "The Logic of 'Being Informed' Revisited and Revised." *Philosophical Studies.*

Barwise, Jon, and Jerry Seligman. 1997. *Information Flow: The Logic of Distributed Systems.* Cambridge: Cambridge University Press.

Bynum, Terrell Ward, and James Moor, eds. 1998. *The Digital Phoenix: How Computers Are Changing Philosophy.* Oxford: Blackwell.

Ess, Charles. 2008. "Luciano Floridi's Philosophy of Information and Information Ethics: Critical Reflections and the State of the Art." *Ethics and Information Technology* 10, no. 2:89–96.

Floridi, Luciano. 1996. *Scepticism and the Foundation of Epistemology: A Study in the Metalogical Fallacies.* Leiden: Brill.

———. 1999. "Information Ethics: On the Philosophical Foundations of Computer Ethics." *Ethics and Information Technology* 1, no. 1:33–52.

———. 2002. "What Is the Philosophy of Information?" *Metaphilosophy* 33, nos. 1–2:123–45. Reprinted in James H. Moor and Terrell Ward Bynum, eds., *CyberPhilosophy: The Intersection of Philosophy and Computing* (Oxford: Blackwell, 2003).

———. 2004. "Outline of a Theory of Strongly Semantic Information." *Minds and Machines* 14, no. 2:197–222.

———. 2005a. "Consciousness, Agents and the Knowledge Game." *Minds and Machines* 15, nos. 3–4:415–44.

———. 2005b. "Is Semantic Information Meaningful Data?" *Philosophy and Phenomenological Research* 70, no. 2:351–70.

———. 2006. "The Logic of Being Informed." *Logique et Analyse* 49, no. 196:433–60.

———. 2007. "In Defence of the Veridical Nature of Semantic Informa-tion." *European Journal of Analytic Philosophy* 3, no. 1:1–18.

———. 2008a. "Information Ethics: A Reappraisal." *Ethics and Informa-tion Technology* 10, nos. 2–3:189–204 (special issue entitled "Luciano Floridi's Philosophy of Information and Information Ethics: Critical Reflections and the State of the Art," edited by Charles Ess).

———. 2008b. "Understanding Epistemic Relevance." *Erkenntnis* 69, no. 1:69–92.

———. 2010a. *Information: A Very Short Introduction.* Oxford: Oxford University Press.

———. 2010b. *The Philosophy of Information.* Oxford: Oxford University Press.

———. Forthcoming. "Information, Possible Worlds, and the Cooptation of Scepticism." *Synthese.*

———. Forthcoming 2. "Semantic Information and the Correctness Theory of Truth."

Floridi, Luciano. ed. 2003. *Erkenntnis. The Blackwell Guide to the Philosophy of Computing and Information.* Oxford: Blackwell.

French, Steven. Forthcoming. "The Interdependence of Structure, Objects and Dependence." *Synthese.*

Hongladarom, Soraj. 2008. "Floridi and Spinoza on Global Information Ethics." *Ethics and Information Technology* 10, no. 2:175–87.

Schumpeter, J. A. 1943. *Capitalism, Socialism, and Democracy.* London: G. Allen and Unwin.

Wilczek, Frank. 2008. *The Lightness of Being: Mass, Ether, and the Unification of Forces.* New York: Basic Books.

11

PHILOSOPHY IN THE INFORMATION AGE

TERRELL WARD BYNUM

Today it is often said that we are living in the "Information Age" and are experiencing an "Information Revolution." The import of such pronouncements, however, is sometimes very unclear. What are the specific characteristics of the Information Age, and when did the Information Revolution begin? Drawing a line at any particular point may seem rather arbitrary; but, for reasons explained below, I focus in this chapter on technical and scientific developments in the 1940s that enabled a remarkable "explosion" of new products and innovations in the 1950s and beyond. The relevant technical and scientific innovations included the creation of electronic computers, the development of a scientific theory of information, and the birth of cybernetics. These mark the beginning of the Information Age, as I understand it, and they generated an enormous number of social and ethical changes and challenges. The Information Revolution, in my view, is *the exponentially growing number of social changes and challenges enabled by electronic technology, the scientific study of information, and the birth of cybernetics.*

Like other scientific and technological revolutions (for example, the Copernican Revolution and the Industrial Revolution), the Information Revolution is having profound effects, not only upon society in general, but also more specifically upon Philosophy. Like a phoenix, Philosophy regularly renews itself, and the Information Revolution seems to be spurring such a renewal. In recognition of this fact, James Moor and I published a book in 1998 entitled *The Digital Phoenix: How Computers Are Changing Philosophy.* There, we noted that computing and related technologies are providing philosophy with "new and evolving subject matters, methods, and models for philosophical inquiry.... Most importantly, computing is changing the way philosophers understand foundational concepts in philosophy, such as mind, consciousness, experience, reasoning, knowledge, truth, ethics and creativity. This trend in philosophical inquiry that incorporates computing in terms of a subject matter, a method, or a model has been gaining momentum steadily. A Digital Phoenix is rising!" (Bynum and Moor 1998, 1). Today, more than a decade later, it is clear that *the Digital Phoenix has indeed risen!* Thus,

Philosophy already has been profoundly influenced by the Information Revolution; and the result has been the creation, in universities around the globe, of new courses, professorships, journals, conferences, scholarships, research centers, grants, and awards.

Two Philosophers of the Information Age

The primary focus of the present chapter is the interaction of Philosophy with the Information Revolution. The results of that interaction already have been so broad and deep that it is impossible to do more, here, than to take note of a few important developments. Consequently, in this chapter, I limit my remarks to comments on the work of two significant philosophers of the Information Age—Norbert Wiener and Luciano Floridi (for additional relevant details about Wiener, see Bynum 2000, 2004, 2005, 2008). Wiener was one of the "founding fathers" of the Information Revolution, and he played a remarkable role in it. He simultaneously helped to create the relevant science and technology; and, with impressive philosophical insight, he anticipated many of the resulting social and ethical consequences. Decades later, equipped with an array of new tools and examples from computer science, systems theory, logic, linguistics, semantics, artificial intelligence, philosophy of mind, philosophy of science, and theoretical physics, Luciano Floridi has spearheaded an ambitious program—his "Philosophy of Information" project—to place the concept of information into the bedrock of Philosophy.

In the pages below, I summarize and compare key ideas of Wiener and Floridi with regard to human nature, the nature of society, the nature of artificial agents (such as robots and softbots), and even the nature of the universe. In the past, new ideas like these have had a powerful impact upon Philosophy, and the Information Revolution is no exception.

Wiener's Role in the Birth of the Information Revolution

As a young man, Wiener was a child prodigy who graduated from Tufts at the age of fourteen and earned a Harvard Ph.D. in Philosophy at eighteen. His doctoral dissertation was on the topic of mathematical logic, and a few years later he became a mathematics teacher at MIT. In 1933, he won the prestigious Bôcher Memorial Prize for developing mathematics that can be used to analyze and describe the seemingly erratic "Brownian motion" of particles in a gas. During World War II, in part because of his knowledge of the mathematics of Brownian motion, Wiener was asked to head a team of scientists and engineers to create a new kind of antiaircraft cannon. Airplanes, by that time, had become so fast and maneuverable that shooting them down was much more difficult than it had been in the past. Wiener and his team decided to design a cannon that would use the new technologies on which they and their colleagues at MIT were working—*radar* to "perceive" and track an airplane, plus *electronic computers* that, without human intervention, could calculate the likely trajectory of an airplane, aim the cannon, and even fire the shell. In the midst of this wartime project, Wiener realized that the new cannon he and his team were designing represented *a revolutionary kind of new machine*, one that would be able to

- gather information about the world,
- store, process and retrieve that information,
- make decisions based upon information processing, and
- carry out those decisions by itself, without human intervention.

Given this project, Wiener realized that after the war the new applied science he and his colleagues were developing could have significant social and ethical consequences. "The choice between good and evil," he said, "knocks at our door."

After the war, Wiener focused much of his attention and energy on the social and ethical implications of his new science, which he had named "cybernetics." In 1948, for example, he published the book *Cybernetics: Or Control and Communication in the Animal and the Machine*; there he described not only the main ideas of his new science but also some social and ethical implications. After *Cybernetics* was published, some of Wiener's friends encouraged him to explain the social implications in even more detail. He took their advice and in 1950 published a book entitled *The Human Use of Human Beings*. Also, in conference presentations, public lectures, and newspaper interviews he explored social and ethical issues that could arise because of cybernetics and electronic computers. With remarkable foresight, he predicted many features of today's Information Age; and near the end of his life in 1964, he wrote a third relevant book, *God & Golem, Inc.: A Comment on Certain Points Where Cybernetics Impinges on Religion*.

In his books, lectures, and interviews after the war, Wiener did not present himself as a philosopher intent upon making new contributions to Philosophy—even though that was one of the things he was doing at the time. He wrote, instead, as a scientist concerned about the social impacts of his scientific achievements and those of his colleagues. Nevertheless, as a young man he had earned a Ph.D. in Philosophy, and he had studied philosophy with influential scholars, such as Josiah Royce, George Santayana, Bertrand Russell, G. E. Moore, Edmund Husserl, and John Dewey. Today, with the advantage of hindsight, we can see that Norbert Wiener was not only a major player in the creation of the Information Revolution, he also laid down a powerful foundation for a new branch of Philosophy that currently is called "Information and Computer Ethics." In addition, he provided—apparently without planning to do so—a new interpretation of human nature, the nature of society, the nature of artificial agents like robots, and even the nature of the universe.

Wiener on the Nature of the Universe

In 1947, Wiener made an important discovery regarding the fundamental role of information in the universe. While dealing with issues in an area that later would be called "information theory," he told some graduate students and colleagues that "information is entropy" (Rheingold 2000, ch. 5)—or more accurately that *entropy is a measure of information* that is contained in every physical being and "lost" in virtually every physical change ("lost" in the sense that it becomes unavailable to form new physical entities). In addition, in the manuscript of *Cybernetics*, which was circulating among his scientific colleagues at the time, Wiener noted that *information is physical, but not matter or energy*. Thus, while describing thinking as information processing in the brain, Wiener wrote that the brain "does not secrete thought 'as the liver does bile,' as the earlier materialists claimed, nor does it put it out in the form of energy, as the muscle puts out its activity. Information is information, not matter or energy. No materialism which does not admit this can survive at the present day" (Wiener 1948, 155). The quantity of physical information "lost" in virtually every physical change is determined by the second law of thermodynamics. Such information is sometimes called "Shannon information," named after Claude Shannon, a colleague of Wiener's. Shannon and Wiener discovered—apparently simultaneously—that entropy is a measure of physical information. Shortly thereafter, Shannon developed the mathematical theory of information and communication for which he became famous. Some physical information is analogue and some is digital, as illustrated by the fact that radio, television, and telephone signals can be either; and some computers are analogue while others are digital. Both analogue information and digital information are governed by the second law of thermodynamics. (Wiener did not address the metaphysical question of which kind of information is the "ultimate stuff" of the universe.)

On Wiener's view, matter-energy and physical information are different physical phenomena, although neither exists without the other. Thus, physical objects and processes actually consist of patterns of information encoded within an ever-changing flux of matter-energy. Every physical object or process, consequently, participates in a creative "coming-to-be" and a destructive "fading away," as old patterns of information erode and new patterns emerge. The discovery by Wiener and Shannon that entropy is a measure of information provided a new way to understand the nature of physical objects and processes. Using today's language, one can say that all physical entities in the universe are "information objects" or "information processes"—a view of the nature of the universe worthy of the Information Age! On this view, even living things are information objects. They store physical information in their genes and use it to create the building blocks of life, like DNA, RNA, proteins, and amino acids. Nervous systems of animals take in, store, and process physical information, thereby making motion, perception, emotion, and thinking possible.

In summary, then, Wiener's account of the nature of the universe is that it consists of information objects and processes. Physical changes in the universe, including the ultimate fading away of a physical object or process, results from an irreversible loss of physical information—an increase in entropy—governed by the second law of thermodynamics. Given that law, it follows that essentially all physical changes decrease available information in the universe, and so every object or process that ever comes into existence will eventually be destroyed. This includes whatever a person might value, like life, wealth, and happiness; magnificent works of art; great architectural structures; cities, societies, civilizations—even the sun, moon, and stars! All are subject to ultimate decay and destruction, because every physical thing in the universe is subject to the second law of thermodynamics.

For this reason, Wiener considered entropy to be *the greatest natural evil.* In doing so he used the traditional distinction between "natural evil," caused by the forces of nature (for example, earthquakes, volcanoes, diseases, floods, tornados, and physical decay), and "moral evil" (for example, human-caused death, injury, and pain). The ultimate natural evil, then, according to Wiener is *entropy*—the loss of available physical information.

Digital Physics

Wiener's view of the nature of the universe anticipated later research and discoveries in physics. In recent times, for example, some "digital physicists"—beginning with Princeton's John Wheeler in 1990 (Wheeler 1990)—have been developing a "theory of everything" which assumes that the universe is fundamentally informational—that every object or entity is, in reality, a pattern of digital information encoded in matter-energy. Wheeler called his hypothesis "it from bit," although the "bits" in question are actually *quantum bits* ("qubits"), which, given the laws of quantum mechanics, can be both extended *and* discrete, positive *and* negative at the same time. Wheeler's "it from bit" hypothesis has been studied and furthered by other scientists in recent years. Their findings reinforce Wiener's view that the universe consists of informational objects and processes interacting with each other. (However, it is important to keep in mind that Wiener did not address himself to the question of whether the information in question is digital or analogue.) According to MIT professor Seth Lloyd,

> The universe is the biggest thing there is and the bit is the smallest possible chunk of information. The universe is made of bits. Every molecule, atom and elementary particle registers bits of information. Every interaction between

those pieces of the universe processes that information by altering those bits. (2006, 3)

. . .

I suggest thinking about the world not simply as a machine, but as *a machine that processes information*. In this paradigm, there are two primary quantities, energy and information, standing on an equal footing and playing off each other. (169)

Contemporary science writer Charles Seife has noted that, indeed, every physical thing in the universe is made of information that obeys the laws of physics:

Information is not just an abstract concept, and it is not just facts or figures, dates or names. It is a concrete property of matter and energy that is quantifiable and measurable. It is every bit as real as the weight of a chunk of lead or the energy stored in an atomic warhead, and just like mass and energy, information is subject to a set of physical laws that dictate how it can behave—how information can be manipulated, transferred, duplicated, erased, or destroyed. And everything in the universe must obey the laws of information, because everything in the universe is shaped by the information it contains. (2006, 2)

. . .

Each creature on earth is a creature of information; information sits at the center of our cells, and information rattles around in our brains.... Every particle in the universe, every electron, every atom, every particle not yet discovered, is packed with information ... that can be transferred, processed, and dissipated. Each star in the universe, each one of the countless galaxies in the heavens, is packed full of information, information that can escape and travel. That information is always flowing, moving from place to place, spreading throughout the cosmos. (3)

Although Wiener's view of the nature of the universe does not require that the information out of which all physical things are composed must be digital, there is currently some evidence that it may indeed be digital. Thus, one of Wheeler's past students, Jacob Beckenstein, discovered the so-called Beckenstein Bound, which sets an upper limit on the amount of physical information that can be contained in a given volume of space. The maximum number of information units (qubits) that can fit within a specific volume is *fixed by the area of the boundary enclosing that space*— one qubit per four "Planck squares" of area (Beckenstein 2003). So, the physical information composing the existing entities of the universe appears to be finite and digital; and only so much physical information can be contained within a specific volume of space. (For an alternative nondigital view, see Floridi 2008a.)

Human Nature

Human beings can be viewed as fundamentally informational, just like other physical entities in the universe, including other animals. So a human is essentially a *pattern* of physical information, which endures over time, in spite of the constant exchange of molecules that occurs through biological metabolism. Thus, Wiener says of human beings,

> We are but whirlpools in a river of ever-flowing water. We are not stuff that abides, but patterns that perpetuate themselves. (1954, 96)

> . . .

> The individuality of the body is that of a flame . . . of a form rather than of a bit of substance. (1954, 102)

Because of breathing, eating, perspiring, digesting, and other metabolic processes, the matter within a person's body is constantly being exchanged with molecules and atoms from outside the body. Nevertheless, the pattern of information that is encoded within a person's body remains similar over time, changing only very gradually. That persisting informational pattern preserves a person's life, functionality, and identity for an extended period of time. Eventually, of course, the pattern changes significantly, and the inevitable results are increasing disability and death—the ultimate destruction of the information pattern that constitutes one's being.

The informational nature of a person makes it possible for him or her to interact with other informational entities in the surrounding environment. In *The Human Use of Human Beings*, Wiener said this:

> Information is a name for the content of what is exchanged with the outer world as we adjust to it, and make our adjustment felt upon it. The process of receiving and of using information is the process of our adjusting to the contingencies of the outer environment, and of our living effectively within that environment. The needs and the complexity of modern life make greater demands on this process of information than ever before. . . . To live effectively is to live with adequate information. Thus, communication and control belong to the essence of man's inner life, even as they belong to his life in society. (1954, 17–18)

Like other animals, human beings are capable of processing physical information within their bodies. The physical structure of any given animal (including a person), according to Wiener, determines the nature and complexity of the information processing in which that animal can engage. For human beings, Wiener emphasized *the tremendous potential for learning and creative action* made possible by human physiology, and he often drew a contrast with other animals, such as insects:

I wish to show that the human individual, capable of vast learning and study, which may occupy about half of his life, is physically equipped, as the ant is not, for this capacity. Variety and possibility are inherent in the human sensorium—and indeed are the key to man's most noble flights—because variety and possibility belong to the very structure of the human organism. (1954, 51–52)

. . .

Cybernetics takes the view that the structure of the machine or of the organism is an index of the performance that may be expected from it. The fact that the mechanical rigidity of the insect is such as to limit its intelligence while the mechanical fluidity of the human being provides for his almost indefinite intellectual expansion is highly relevant to the point of view of this book. (1954, 57; italics in the original)

. . .

[M]an's advantage over the rest of nature is that he has the physiological and hence the intellectual equipment to adapt himself to radical changes in his environment. The human species is strong only insofar as it takes advantage of the innate, adaptive, learning faculties that its physiological structure makes possible. (1954, 58)

According to Wiener, there is a fundamental relationship between the purpose of a human life and the kind of internal information processing that occurs within a human body. Wiener considered *flourishing as a person* to be the overall *purpose* of life—flourishing in the sense of realizing one's full human potential in variety and possibility of choice and action. To flourish, a person must engage in a wide range of information processing activities, such as perceiving, organizing, remembering, inferring, deciding, planning, acting, and so forth. It follows that human flourishing is utterly dependent upon information processing.

Artificial Agents

Beginning with his book *Cybernetics*, Wiener described human beings (as well as other animals) as *dynamic information processing systems* with component parts that communicate with each other internally by means of feedback loops. Such internal communications unify animals (including humans) so that all their different parts can work together toward common goals. In *Cybernetics*, Wiener assumed that there will be machines that function in a similar manner. Some machines, he said, will make decisions and carry them out by themselves, while others will even learn and adjust their future behavior to take account of their past.

Wiener worried about the possibility that machines that learn and make decisions might generate significant ethical risks. Indeed, he was especially concerned about the possibility that someone could foolishly create "artificial agents" (as we would call them today) that humans may not be able to control—agents that could act on the basis of values that humans do not share. Consequently, Wiener cautioned that a prudent man

> will not leap in where angels fear to tread, unless he is prepared to accept the punishment of the fallen angels. Neither will he calmly transfer to the machine made in his own image the responsibility for his choice of good and evil, without continuing to accept a full responsibility for that choice. (1950, 211–12)

> . . .

> [T]he machine . . . which can learn and can make decisions on the basis of its learning, will in no way be obliged to make such decisions as we should have made, or will be acceptable to us. For the man who is not aware of this, to throw the problem of his responsibility on the machine, whether it can learn or not, is to cast his responsibility to the winds, and to find it coming back seated on the whirlwind. (1950, 212)

To prevent this kind of disaster, according to Wiener, the world will need ethical rules for artificial agents.

In 1950, in *The Human Use of Human Beings*, Wiener predicted that machines *will join humans as active participants in society*. Some machines, he said, eventually will participate, along with humans, in the activity of creating, sending, and receiving messages that function as the "cement" that binds society together: "It is the thesis of this book that society can only be understood through a study of the messages and the communication facilities which belong to it; and that in the future development of these messages and communication facilities, messages between man and machines, between machines and man, and between machine and machine, are destined to play an ever-increasing part" (1950, 9).

Wiener predicted that in the future there will be digital computers with robotic appendages. Such robots will participate in the workplace and replace thousands of human factory workers, both blue collar and white collar. Wiener also foresaw artificial limbs and other human-created body parts—cybernetic "prostheses"—that will be *merged with human bodies* in order to help persons with disabilities. Such devices, he said, could even be used to endow able-bodied persons with unprecedented powers. Thus, Wiener foresaw societies in which cyborgs (as we would call them today) would play a significant role. Such societies would need to establish ethical policies to govern cyborg behavior.

With these things in mind, Wiener foresaw a "Machine Age" or "Automatic Age" where machines will be integrated into society. They will create, send, and receive messages; gather information; make decisions; take actions; reproduce themselves; and even be merged with human bodies to create beings with vast new powers. These predictions were not just speculations, because Wiener had already witnessed or designed early versions of game-playing machines (checkers, chess, war, business), artificial hands with motors to be controlled by a person's brain, and self-reproducing nonlinear transducers. (See especially Wiener 1964.)

Given Wiener's predictions about such machines, people began asking him whether those machines would be "alive." He responded by saying that such questions could be considered semantic quibbles, rather than truly scientific questions: "Now that certain analogies of behavior are being observed between the machine and the living organism, the problem as to whether the machine is alive or not is, for our purposes, semantic and we are at liberty to answer it one way or the other as best suits our convenience" (Wiener 1954, 32). Wiener *did* believe that questions about the "intellectual capacities" of machines, when appropriately formulated, could be genuine scientific questions: "*Cybernetics takes the view that the structure of the machine or of the organism is an index of the performance that may be expected from it. . . .* Theoretically, if we could build a machine whose mechanical structure duplicated human physiology, then we could have a machine whose intellectual capacities would duplicate those of human beings" (1954, 57; italics in the original). In 1964 (in *God & Golem, Inc.*) Wiener was skeptical of the idea that the physical structure of a machine would ever duplicate that of a human brain. Electronic components in those days were simply too large and generated too much heat to be crammed together like neurons in a human brain. (Given today's nanocircuits, perhaps Wiener would be less skeptical.)

By treating animals (including humans) and cybernetic machines as dynamic information-processing systems, Wiener began to see traditional differences between mechanism and vitalism, living and nonliving, human and machine as blurry pragmatic distinctions, rather than unscalable metaphysical "walls."

Wiener on Society in the Information Age

Wiener considered communities and societies to be *second-order cybernetic systems*, because their members are themselves cybernetic systems: "It is certainly true that the social system is an organization like the individual; that it is bound together by a system of communication; and that it has a dynamics, in which circular processes of a feedback nature play an important part" (1948, 33). This is true even of bee hives, ant colonies, and herds of mammals, not just human communities. Information processing and the flow of information are keys to their nature and their successful functioning. For this reason, Wiener noted that communication is "the central phenomenon of society" (1950, 229). As a consequence, Wiener's discussions of the nature of society often included consideration of communication networks and their roles in the community. Late in Wiener's life, a crude worldwide telecommunications net-work—consisting of telephone, telegraph, cable, and radio facilities—already existed. Thus, even though Wiener died several years before the creation of the Internet, he nevertheless identified, in the 1950s and early 1960s, some ethical questions that are commonly associated with today's Internet. For example, one of Wiener's thought experiments was about people working on a job by using long-distance telecommunication facilities (today's "teleworking" or "telecommuting"). He illustrated this possibility by imagining an architect in Europe who could guide the construction of a building in America while remaining in Europe. The imagined architect made use of telegrams, telephone messages, and an early form of faxing called "Ultrafax" to send and receive instructions, plans, and photographs (Wiener 1950, 104–5, and 1954, 98).

Another telecommunications possibility that Wiener briefly discussed was "virtual communities" (as we would call them today). Thus, in 1948, Wiener noted that "[p]roperly speaking, the community extends only so far as there extends an effectual transmission of information" (1948, 184) And in 1954, he noted: "Where a man's word goes, and where his power of perception goes, to that point his control and in a sense his physical existence is extended. To see and to give commands to the whole world is almost the same as being everywhere. . . . Even now the transportation of messages serves to forward an extension of man's senses and his capabilities of action from one end of the world to another" (1954, 97–98). Wiener clearly saw that long-distance telecommunication facilities would someday create significant possibilities for people to cooperate "virtually" (as we would say today), either on the job, or as members of groups and communities, or even as citizens participating in government. (See Wiener on world government in Wiener 1954, 92.) Given Wiener's assumption that exchanges of messages constitute "cement that holds society together," today's world must be rapidly "morphing" into a global society. People around the globe are exchanging billions of messages daily, using e-mail, "texting," instant messages, blogging, "tweeting," video posting, and on and on.

The Philosophy of Information: Floridi's Ambitious Project

In the early days of the Information Revolution, Wiener was busy helping to lay its foundations. In addition, he predicted, with remarkable insight, an impressive number of its future consequences. By the mid-1990s, a significant number of Wiener's predictions had already come true—the invention of numerous cybernetic artifacts, the replacement of many human workers by robots, telecommuting of workers from home, the appearance of "virtual communities" (as we would call them today), and the creation of new ethical challenges (Wiener's "choices between good and evil")—to cite only a few examples. In this context, philosopher

Luciano Floridi launched his ambitious Philosophy of Information project to create a new philosophical paradigm. Floridi considered a number of existing paradigms—like analytic philosophy, phenomenology, and existentialism—to be "scholastic," that is, *stagnant* as intellectual enterprises:

> Scholasticism, understood as an intellectual topology rather than a scholarly category, represents the inborn inertia of a conceptual system, when not its rampant resistance to innovation. It is *institutionalized philosophy* at its worst. ... It manifests itself as a pedantic and often intolerant adherence to some discourse (teachings, methods, values, viewpoints, canons of authors, positions, theories, or selections of problems, etc.), set by a particular group (a philosopher, a school of thought, a movement, a trend, etc.), at the expense of alternatives, which are ignored or opposed. (2002, 125)

According to Floridi, Philosophy "can flourish only by constantly re-engineering itself. A philosophy that is not timely but timeless is not an impossible *philosophia perennis*, which claims universal validity over past and future intellectual positions, but a stagnant philosophy" (2002, 128). Floridi set for himself—as an alternative to scholastic philosophical systems—the ambitious task of creating a new philosophical paradigm that can become part of the "bedrock" of Philosophy (*philosophia prima*). At the heart of that project was to be the concept of *information*, a concept with multiple meanings, and also "a concept as fundamental and important as being, knowledge, life, intelligence, meaning, or good and evil—all pivotal concepts with which it is interdependent—and so equally worthy of autonomous investigation. It is also a more impoverished concept, in terms of which the others can be expressed and interrelated, when not defined" (2002, 134).

To carry out his project, Floridi had available to him many new methods and conceptual resources—developed after Wiener's pioneering days—in fields such as logic, computer science, systems theory, artificial intelligence, philosophy of mind, linguistics, semantics, philosophy of science, and theoretical physics. Since the mid-1990s, Floridi has used these powerful new resources to advance his Philosophy of Information project, and a number of graduate students and philosophical colleagues have joined him in that effort. His project has grown and matured into a broad research program addressing a substantial array of philosophical issues. These range from the (deceptively simple but very complex) question of the nature of information to such topics as the informational nature of the universe, the semantics of scientific models, symbol grounding and consciousness, the nature and ethics of artificial agents, the foundation and uniqueness of computer ethics, the nature and role of artificial companions in a human life, the role of information in reasoning and logic, and many more. In the limited space of the present chapter, it is possible to cover only a few relevant topics; so the sections below focus only upon Floridi's views about the nature of the universe, human nature, artificial agents, and the nature of society. At the end, some key ideas of Floridi and Wiener are compared.

Floridi on the Nature and Goodness of the Universe

Methodologically, Floridi is a "constructionist" who adopts the view that ultimate reality (Kant's "noumenal" world of "things-in-themselves") is unknowable—a "black box" into which one can never see. For this reason, even though ultimate reality provides certain affordances and imposes certain constraints upon our experiences, we can never know why and how. In a quest to understand things-in-themselves, the best that we can do is to construct *models* of ultimate reality (or, perhaps, *models of parts* of ultimate reality). Knowledge, truth, and semantics, according to Floridi, apply to our models, not to ultimate reality, which remains forever inaccessible.

The world that we experience (Kant's phenomenal world) is the sum total of our models of reality. It follows that we live in a different world when we significantly change the objects and/or processes within our models. This is *not* a version of relativism, however, because we can compare models with regard to their ability to accommodate the affordances and constraints of unknowable ultimate reality. With regard to *semantic* information, according to Floridi, it must be "well-formed, meaningful, and *true*." So-called false information, he argues, is not information at all, it is merely *misinformation*. Even if it is well formed and meaningful, it cannot be information at all, because genuine information is true.

Floridian models are constructed using a "method of abstraction," which he and his colleague J. W. Sanders developed by adapting Formal Methods from computer science. Their philosophical method involves the selection of a set of "observables" at a given "level of abstraction." By attributing certain "behaviors" to the observables, one builds a model of the entity being analyzed, and this can be tested against our experiences, observations, and experiments. The best models are those which most successfully achieve "informativeness, coherence, elegance, explanatory power, consistency, predictive power, etc." (Floridi 2004, 8) and allow successful interactions with the world. (See Floridi 2010, ch. 8.)

In his 2004 article "Informational Realism" (see also Floridi 2008a) Floridi argues that, at a certain level of abstraction, *all objects in the universe are data structures* composed of "mind-independent points of lack of uniformity." These are Platonic in nature, rather than physical data, so they do not obey the laws of physics. Thus, "the outcome is *informational realism*, the view that the world is the totality of informational objects dynamically interacting with each other" (Floridi 2004, 7; italics in the original). In summary, then, at the informational level of abstraction, every existing entity is a "data structure"—an "informational object"—composed of Platonic "relations" describable as "mind-independent points of lack of uniformity." (See especially Floridi 2008a.)

INFORMATION ETHICS

According to Floridi, in addition to being composed of informational objects, the *universe is fundamentally good*, and that goodness is *not dependent upon human ethical judgments*. This is a major metaphysical assumption of Floridi's "macroethics" (his term), which he calls INFORMATION ETHICS. (SMALL CAPITALS are used here to distinguish Floridi's theory from the more general field of information ethics in the broadest sense.) His theory is similar, in one way, to traditional ethical theories like virtue ethics, deontologism, consequentialism, and contractualism, because it is intended to be applicable to all ethical situations. On the other hand, INFORMATION ETHICS is also *different* from these traditional theories because it is intended to *supplement* them with further ethical considerations, rather than to replace them; and it also avoids an anthropocentric focus upon human actions, characters, and values (Floridi 2005). Thus, INFORMATION ETHICS considerations, in a given circumstance, can be perfectly consistent with traditional ethical judgments; but it is also possible, in some circumstances, for INFORMATION ETHICS considerations to be overridden by more traditional ethical concerns.

According to Floridi, then, every existing entity in the universe, when viewed from a certain level of abstraction, is an informational object, and each such object has a characteristic data structure that constitutes its very nature. As a result, he refers to the universe as "the infosphere." Every entity in the infosphere can be destroyed or damaged by altering its characteristic data structure, thereby preventing it from "flourishing." Even if one puts aside all *traditional* anthropocentric ethical considerations—from theories such as deontologism, utilitarianism, contractualism, and virtue theory—every existing entity in the infosphere, viewed from the informational level of abstraction, would still have at least a minimum of ethical worth because the universe is basically good, and information itself (understood not semantically but in terms of data structures) has at least a minimum worth. Viewed from the informational level of abstraction, therefore, damaging the data structure of an informational object, when there are no overriding ethical considerations that could be advanced from traditional anthropocentric theories, would amount to unjustified "impoverishment of the infosphere." Floridi calls such damage or destruction "entropy." Although he borrowed this term from physics, he does *not* mean thermodynamic entropy, which is subject to the laws of physics. Instead, Floridian entropy is unjustified impoverishment of the infosphere, and it should be avoided or minimized. With this in mind, he introduced the "fundamental principles" of INFORMATION ETHICS:

1. entropy ought not to be caused in the infosphere (null law)
2. entropy ought to be prevented in the infosphere
3. entropy ought to be removed from the infosphere
4. the flourishing of informational entities as well as the whole infosphere ought to be promoted by preserving, cultivating and enriching their properties

INFORMATION ETHICS, because it views every existing entity as an informational object with at least a minimal moral worth, shifts the focus of ethical consideration *away from* actions, characters, and values of human agents, and *toward* the "evil" (damage, dissolution, destruction) suffered by objects in the infosphere. Given this approach, every existing entity—humans, other animals, organizations, plants, nonliving artifacts, digital objects in cyberspace, pieces of intellectual property, stones, Platonic abstractions, possible beings, vanished civilizations—all can be interpreted as *potential agents* that affect other entities, and as *potential patients* that are affected by other entities. Thus, Floridi's INFORMATION ETHICS can be described as a "patient-centered" nonanthropocentric ethical theory instead of the traditional "agent-centered" anthropocentric theories.

Some critics of Floridi's INFORMATION ETHICS have claimed that his metaphysical presupposition about the inherent goodness of the universe is unnecessary and unjustified. In reply, Floridi has said:

> The actual issue is whether Goodness and Being (capitals meant) might be two sides of the same concept, as Evil and Non-Being might be. . . . [T]he reader sufficiently acquainted with the history of Western philosophy need not be told about classic thinkers, including Plato, Aristotle, Plotinus, Augustine, Aquinas and Spinoza, who have elaborated and defended in various ways this fundamental equation. For Plato, for example, Goodness and Being are intimately connected. Plato's universe is value-ridden at its very roots: value is there from the start, not imposed upon it by a rather late-coming, new mammalian species of animals, as if before evolution had the chance of hitting upon *homo sapiens* the universe were a value-neutral reality, devoid of any moral worth. (2008c)

According to Floridi, seeing something in a particular way—that is, adopting a particular level of abstraction in order to model it—always has a purpose. If that purpose is fulfilled well and fruitfully, then one is justified in taking that perspective. In this case, by viewing the universe as inherently good, consisting of informational objects, their relationships and processes, Floridi is able to accomplish at least three major things:

1. Make sense of the awe and respect that one feels when contemplating or experiencing the vast and beautiful universe (like Toaism, Buddhism, Platonism, Aristotelianism, Stocism, Spinozian philosophy, and so on).

2. Develop a way to apply moral reasoning and accountability to the rapidly increasing number of artificial environments and agents (robots, webbots, cyborgs, virtual communities, and so forth) that are being created in our midst by the billions. (See below.)
3. Develop a way to understand *distributed moral accountability* within complex social agents like corporations, organizations, virtual communities, governments, and so on. (See below.)

Artificial Agents and Ethics

When Wiener first predicted, in the late 1940s, that there would be machines that make decisions, including even some that learn from their past "experiences," many people considered his predictions to be either science fiction or major exaggerations. Today, Wiener's predictions have been fulfilled overwhelmingly, and so they now seem to be *understatements*. Some of today's "artificial agents" are robotic hardware devices like those that Wiener envisioned, but there also are others, such as software agents like webbots that "crawl" through computer networks, and softbots that reside in iPods, laptops, cell phones—even in home appliances, digital cameras, hearing aids, and so on. Today artificial agents perform many tasks: correct spelling, delete "spam" e-mail, find and remove computer viruses, control nuclear power plants, fly airplanes, control rail networks for high-speed trains, launch missiles, inject patients with medicine, perform delicate surgery on living human beings, and on, and on. Thus Wiener certainly was right when he predicted that someday there would be a society that would need ethical rules and procedures to govern artificial agents. Our own society fits that description today. In several articles, Floridi and Sanders (1999, 2001, 2004) addressed this need; and in doing so, they sought to achieve three goals (Floridi and Sanders 2004):

1. Provide "an effective characterization" of an agent.
2. Provide an appropriate account of good and evil that artificial agents are capable of bringing about.
3. Provide an explanation of how and why to hold artificial agents morally accountable, even if they are "mindless" and thus without mental states.

Because a human being is a paradigm example of an agent, Floridi and Sanders's characterization of an agent needs to fit humans. In addition, it also must fit softbots, robots, and other artificial agents, such as virtual communities, organizations, corporations, and governments. The characterization that Floridi and Sanders developed includes three criteria that an entity must meet to be an agent (Floridi and Sanders 2004):

i. *Interactivity*: The agent and its environment can act upon each other.

ii. *Autonomy*: The agent is able to change its own state independently of its interactions with the environment. An agent, therefore, must have at least two states and be "decoupled" to some extent from its environment.
iii. *Adaptability*: The agent's interactions with the environment can change the transition rules by which it changes state; that is, the agent's capacity to change its own states can evolve because of its own past interactions. (For humans or animals, we say that they "learn from their experiences.")

According to Floridi and Sanders, to determine whether a given entity is an agent, one must specify *the level of abstraction* at which it is being considered, because something could be viewed as an agent at one level of abstraction but not at a different level. For example, a person certainly *is* an agent, given our everyday understanding of what a human being is, but considered merely as a physical object situated in a specific part of space-time, the person is *not* an agent at that level of abstraction.

Moral Agents

Having provided criteria for being an agent, Floridi and Sanders defined the notion of a "*moral* agent": "An action is said to be morally qualifiable if and only if it can cause moral good or evil. An agent is said to be a moral agent if and only if it is capable of morally qualifiable action" (2004, 364). The term "action" as used by Floridi and Sanders does not presuppose that the agent has mental states like intentions or beliefs, let alone "free will" (in any of the traditional senses of that term). An *action*, for Floridi and Sanders, is just an activity in which an agent causes an effect. Thus, for example, an agent that is a computer "worm" which somehow gets into a computer system of a nuclear power plant, makes and executes a decision, and thereby causes a catastrophe, has engaged in a morally evil action, in spite of the fact that it was totally "mindless," with no intentions or knowledge. Or, to offer a positive example, a mindless computerized medical agent that saves a patient's life by injecting the patient with appropriate medicine in a crisis has engaged in a morally good action.

Moral Responsibility Versus Moral Accountability

Critics of Floridi and Sanders sometimes have argued that it is a mistake to call the activities of mindless beings moral or immoral, because they *cannot be held responsible* for what they do. Floridi and Sanders reply that the objection does not distinguish between holding an agent *accountable* —therefore subject to *censure*—on the one hand, and holding it *responsible*—therefore subject to *blame and praise, punishment and reward*—on the other hand:

Human moral agents who break accepted conventions are censured in various ways of which the main alternatives are: (a) mild social censure with the aim of changing and monitoring behavior; (b) isolation, with similar aims; (c) death. What would be the consequences of our approach for artificial moral agents?

Preserving consistency between human and artificial moral agents [in Cyberspace], we are led to contemplate the following analogous steps for the censure of immoral artificial agents: (a) monitoring and modification (i.e. 'maintenance'); (b) removal to a disconnected component of Cyberspace; (c) deletion from Cyberspace (without backup). (2004, 376)

By distinguishing between accountability and responsibility, according to Floridi and Sanders, we are able to extend ethical considerations to artificial agents like robots and webbots, and we also are able to better understand why it is appropriate to hold young children *accountable* for their behavior, even if they are not yet morally *responsible*. In addition, "[i]t facilitates the discussion of the morality of agents not only in Cyberspace but also in the biosphere—where animals can be considered moral agents ... —and in what we have called contexts of 'distributed morality', where social and legal agents [like corporations, organizations and governments] can now qualify as moral agents. The great advantage is a better grasp of the moral discourse in non-human contexts" (Floridi and Sanders 2004, 377).

Human Nature and the Information Society

At the informational level of abstraction, Floridi views the universe as *the totality of informational objects dynamically interacting with each other*—the infosphere. This includes human beings as well as all other biological organisms; plus all artificial agents; every other physical object; and even "Platonic" entities that are not in physical space-time. In addition, second-order informational entities are included in the infosphere—entities whose parts or members are themselves informational objects; for example, families, organizations, corporations, communities, governments, and whole societies. In summary, according to Floridi, human beings are informational objects dynamically interacting with a world of other informational objects, and societies are complex dynamic second-order informational objects whose members are themselves dynamic informational objects.

Previously, people did not think of themselves as informational objects, nor did they consider most of the objects in their environments—homes, cars, highways, buildings, cookware, and so on—to be dynamic informational objects, even though—at the informational level of abstraction—that is exactly what they are. According to Floridi (2007), *people soon will think of themselves as informational objects*. This will happen, he said, because information and communication technologies are rapidly being incorporated into everyday objects in order to make them interactive with us and with each other. Common, familiar objects, then, soon will be so profoundly reengineered that Floridi has coined the term "reontologisation" to name the reengineering process. Our "reontologized" cookware appliances will communicate with us and with each other as they cook our food. Our refrigerators will learn our dietary preferences and notify us, or the grocery service, when we are running out of certain foods. Our physical belongings will stay in touch with us over the Web when we travel, and many objects in our environment with learn from their "experiences," make decisions, and take actions accordingly as they interact with us and each other.

In summary, the reontologized environment in which we soon will be living, according to Floridi, will consist of humans, artificial agents, and everyday objects, all wirelessly intercommunicating. This "ubiquitous computing" or "ambient intelligence" will make the world seem almost *alive* to us, and today's distinction between being "offline" and being "online" in cyberspace will vanish. In short, we will think of ourselves as *interconnected informational organisms*—"inforgs" (Floridi's term)—who live in a complex and dynamic society of other inforgs, both biological and artificial; and "we shall increasingly feel deprived, excluded, handicapped or poor to the point of paralysis and psychological trauma whenever we are disconnected from the infosphere, like fish out of water" (Floridi 2007, 64). In such a circumstance, the "digital divide" between informationally advantaged societies and "have-not" societies will become a huge social and ethical chasm.

According to Floridi, this remarkable new way of interpreting human nature represents a "fourth revolution" in human self-understanding. It is

> part of a wide and influential informational turn, a fourth revolution in the long process of reassessing humanity's fundamental nature and role in the universe. We are not immobile, at the center of the universe (Copernicus); we are not unnaturally distinct and different from the rest of the animal world (Darwin); we are far from being entirely transparent to ourselves (Freud). We are now slowly accepting the idea that we might be informational organisms among many agents (Turing), ... not so dramatically different from clever, engineered artifacts, sharing with them a global environment that is ultimately made of information. (Floridi 2008b, 654)

Concluding Remarks: Comparing Wiener and Floridi

Even though the Information Age emerged quite recently, it already includes far-reaching technical, scientific, economic, political, social, psychological, and philosophical changes. Because of limited space in this chapter, it has been possible to explore only a handful of related philosophical ideas—namely, those of Norbert Wiener and Luciano Floridi regarding the nature of the universe, human nature, artificial

agents, and society. Wiener was a pioneer who helped to create much of
the science and technology that brought about the Information Revolution. In addition, he had an impressive ability to "see distantly" many of
the resulting social and ethical challenges of today's Information Age.
Several decades after Wiener's great achievements—equipped with new
tools and examples from computer science, systems theory, logic, linguistics, semantics, artificial intelligence, philosophy of mind, philosophy of
science, and theoretical physics—Floridi launched his ambitious Philosophy of Information program to place the concept of information into
the bedrock of philosophy. Wiener's foresight turned out to be remarkably accurate, and Floridi's project appears to be headed for admirable
success.

Initially, when one considers the relevant views of Wiener and Floridi,
they seem to be very similar. Thus, both of them view the universe as
essentially informational—made of dynamically interacting informational objects and processes. Both see human beings as informational
objects. And both say that entropy is a significant evil. First appearances,
though, are deceptive, because Wiener was a materialist and Floridi is a
Platonist, and consequently they interpret the central ideas of entropy and
information very differently.

As explained above, according to Wiener the information of which the
universe is composed is *physical* and *obeys the laws of physics*. It is
syntactical, rather than semantic. It is the kind of information carried
in radio waves, telephone lines, and television cables. It is encoded in
the DNA of every living entity and carried by every subatomic particle.
All physical entities, including human beings, according to Wiener,
are patterns of such information that persist for a time, but gradually evolve, erode, and finally dissipate. Entropy is a measure of that
erosion and dissipation. Wiener did not address the question of
whether the information of which the universe is composed is digital or
analogue.

According to Floridi, the information of which the universe is composed
is nonphysical and therefore does not obey laws of physics, such as the
second law of thermodynamics. It is *Platonic* information—"mind-independent points of lack of uniformity"—that constitute the data structures,
not only of familiar objects like tables and chairs, humans and computers,
but also of nonmaterial Platonic entities such as possible beings, intellectual property, and unwritten stories from vanished civilizations.

Can Wiener and Floridi Be Reconciled?

Compared to that of Floridi, Wiener's philosophy of information is
incomplete. It was an incidental byproduct of his scientific research
projects and social concerns, and thus it was not a carefully chosen
philosophical project like Floridi's. Perhaps it is possible, however, to

reconcile most of Wiener's views with Floridi's. Given Wiener's belief, for example, that the "loss" of available physical information—thermodynamic entropy—is *the greatest evil* in the universe, Wiener could logically accept the idea that *the universe is basically good* to the extent that it continues to contain available physical information for the creation of good objects and processes. Also, nothing in Wiener's philosophy of information would be inconsistent with Floridi's analysis of artificial agents and artificial evil. In addition, Wiener certainly could agree with Floridi's observation that, in today's society, ubiquitous computing is placing more and more objects online, thereby diminishing the distinction between online and offline existence—and turning us all into inforgs.

Since Wiener did not develop a metaphysical theory of the ultimate nature of the informational "stuff" that composes the universe, it would, perhaps, be logically possible for him to accept Floridi's view that such information consists of "mind-independent points of lack of uniformity." However, Wiener was a confirmed materialist while Floridi is a committed Platonist, so for Wiener, the "points of lack of uniformity" would have to be something like the digital physicist's notion of qubits (which are simultaneously discrete and continuous, or positive and negative), rather than Floridi's Platonic "relations" that do not obey physical laws. Given this stark metaphysical difference, the positions of Wiener and Floridi appear to be, in the end, philosophically irreconcilable. A materialist philosopher, perhaps, might some day develop a rival philosophy of information as powerful and well developed as Floridi's Platonic one. In metaphysics, history often repeats itself!

Acknowledgments

I am especially thankful to Luciano Floridi and Patrick Allo for helpful suggestions that improved this chapter significantly. Any shortcomings that remain are my own. I am also thankful to Armen Marsoobian, editor in chief of *Metaphilosophy*, for financial support and a speaking opportunity that benefited the chapter. In addition, I am grateful to Southern Connecticut State University for financial support and a sabbatical leave that helped to make the chapter possible. The present chapter is an expanded and significantly revised version of the Joseph Weizenbaum Address, which I gave at the CEPE2009 conference in Corfu, Greece. I would like to thank the audience there for their very helpful comments and questions.

References

Beckenstein, J. D. 2003. "Information in the Holographic Universe." *Scientific American* (August 2003): 58–65.

Bynum, Terrell Ward. 2000. "The Foundation of Computer Ethics." *Computers and Society* 30, no. 2:6–13.

———. 2004. "Ethical Challenges to Citizens of the 'Automatic Age': Norbert Wiener on the Information Society." *Journal of Information, Communication and Ethics in Society* 2, no. 2:65–74.

———. 2005. "Norbert Wiener's Vision: The Impact of the 'Automatic Age' on Our Moral Lives." In *The Impact of the Internet on Our Moral Lives*, edited by Robert J. Cavalier, 11–25. Albany: State University of New York Press.

———. 2008. "Norbert Wiener and the Rise of Information Ethics." In *Moral Philosophy and Information Technology*, edited by W. J. van den Hoven and John Weckert, 8–25. Cambridge: Cambridge University Press.

Bynum, Terrell Ward, and James H. Moor editors. 1998. *The Digital Phoenix: How Computers Are Changing Philosophy*. Oxford: Blackwell. (Revised edition 2000.)

Floridi, Luciano. 2002. "What Is the Philosophy of Information?" In *Cyberphilosophy: The Intersection of Computing and Philosophy*, edited by James H. Moor and Terrell Ward Bynum, 115–38. Oxford: Blackwell.

———. 2004. "Informational Realism." In *Computers and Philosophy 2003: Selected Papers from the Computers and Philosophy Conference CAP2003*, edited by John Weckert and Yeslam Al-Saggaf, 7–12. Australian Computer Society, Conferences in Research and Practice in Information Technology.

———. 2005. "Information Ethics, Its Nature and Scope." *Computers and Society* 36, no. 3:21–36.

———. 2007. "A Look into the Future Impact of ICT on Our Lives." *Information Society* 23, no. 1:59–64.

———. 2008a. "A Defence of Informational Structural Realism." *Synthese* 161, no. 2:219–53.

———. 2008b. "Artificial Intelligence's New Frontier: Artificial Companions and the Fourth Revolution." *Metaphilosophy* 39, nos. 4–5:651–55.

———. 2008c. "Understanding Information Ethics: Replies to Comments." Reply to John Barker, *American Philosophical Association Newsletter on Philosophy and Computers* 8, no. 2. Available at www.apaonline.org/publications/newsletters/v08n2_Computers_05.aspx (accessed on 1 April 2010).

———. 2010. *The Philosophy of Information*. Oxford: Oxford University Press.

Floridi, Luciano, and J. W. Sanders. 1999. "Entropy as Evil in Information Ethics." In *Etica & Politica* 1, no. 2 (special issue edited by Luciano Floridi). Available at www2.units.it/˜etica/1999_2/index.html (accessed on 1 April 2010).

———. 2001. "Artificial Evil and the Foundation of Computer Ethics." *Ethics and Information Technology* 3, no. 1:55–66.

———. 2004. "On the Morality of Artificial Agents." *Minds and Machines* 14, no. 3:349–79.

Lloyd, Seth. 2006. *Programming the Universe: A Quantum Computer Scientist Takes On the Cosmos.* New York: Knopf.

Rheingold, Howard. 2000. *Tools for Thought.* Cambridge, Mass.: MIT Press. (Originally published in 1985 by Simon and Schuster.)

Seife, Charles. 2006. *Decoding the Universe: How the New Science of Information Is Explaining Everything in the Cosmos, from Our Brains to Black Holes.* New York: Viking Penguin.

Wheeler, John A. 1990. "Information, Physics, Quantum: The Search for Links." In *Complexity, Entropy, and the Physics of Information*, edited by W. H. Zureck. Redwood City, Calif.: Addison Wesley.

Wiener, Norbert. 1948. *Cybernetics: Or Control and Communication in the Animal and the Machine.* New York: Technology Press, John Wiley and Sons.

———. 1950. *The Human Use of Human Beings.* Boston, Mass.: Houghton Mifflin.

———. 1954. *The Human Use of Human Beings.* Second edition revised. Garden City, N.Y.: Doubleday Anchor.

———. 1964. *God & Golem, Inc.: A Comment on Certain Points Where Cybernetics Impinges on Religion.* Boston, Mass.: MIT Press. (Also published in London in the same year by Chapman and Hall.)

INDEX